Dynamic Migration Characteristics of CH_4 During Gas Injection Displacement in Coal Seams

Gongda Wang • Haoran Gong

Dynamic Migration Characteristics of CH4 During Gas Injection Displacement in Coal Seams

Springer

Gongda Wang
University of Science and
Technology Beijing
Beijing, China

Haoran Gong
Henan University of Engineering
Zhengzhou, China

ISBN 978-981-95-8878-7 ISBN 978-981-95-8879-4 (eBook)
https://doi.org/10.1007/978-981-95-8879-4

This Springer imprint is published by the registered company Springer Nature Singapore Pte Ltd.
The registered company address is: 152 Beach Road, #21-01/04 Gateway East, Singapore 189721, Singapore

Competing Interests The authors have no competing interests to declare that are relevant to the content of this manuscript.

Contents

Chapter 1
Introduction

1.1 Overview

According to the BP Statistical Review of World Energy [1], the global proven natural gas reserves are 211 trillion cubic meters, with a total production of 4.25 trillion cubic meters and a total consumption of 4.01 trillion cubic meters, accounting for approximately 24% of the total primary energy consumption. Of this, coalbed methane (CBM or CSG) accounts for 6–9% of the natural gas production. Although its share is relatively small, the development and utilization of coalbed methane, along with its additional functions such as environmental protection [2] and mine gas management [3, 4], play a crucial role in the growth of natural gas production and the prevention of mine gas hazards. Statistics show that, with increased global attention to mine gas management and improvements in gas treatment technologies, the frequency of mine gas accidents and the number of casualties have generally decreased in recent years. However, since 2024, significant gas accidents leading to casualties have occurred in places, such as Pingmei No. 12 Mine in Henan Province, China, and Tabas Mine in South Khorasan Province, Iran. Furthermore, in some high-gas or intensively mined low-gas mines, gas exceeding the safety limits still sometimes affects mining efficiency. This indicates that even today, rapid and efficient gas prevention technologies remain an important research topic for ensuring safe and efficient coal mining and balancing mining operations in China.

Gas drainage enhancement technology is the fundamental approach to mine gas prevention and control. Unlike other gas enhancement techniques, such as protective layer pressure relief [5], hydraulic fracturing [6], hydraulic punching [7], water jetting [8], abrasive jetting [9], and CO_2-induced phase change fracturing [10], which alter the stress state or physical properties of the coal reservoir to increase coal seam fractures and permeability, thereby reducing gas flow resistance, the gas injection displacement technique involves injecting "non-methane" gases into the coal body at pressures higher than the reservoir gas pressure. This technique

G. Wang, H. Gong, *Dynamic Migration Characteristics of CH4 During Gas Injection Displacement in Coal Seams*,
https://doi.org/10.1007/978-981-95-8879-4_1

promotes the desorption of adsorbed gas through competitive adsorption between the injected gases and methane adsorbed on the coal matrix surface, as well as by lowering the methane partial pressure. Additionally, the injected "gas flow," which is higher than the original reservoir pressure, can form a pressure gradient within the reservoir, driving the gas in the coal seam fracture system from a stagnant state to a flowing state, thereby accelerating gas migration within the coal seam fractures. The technical principle of this method is fundamentally different from that of other techniques. Its applicability is relatively unrestricted by the physical properties of the coal reservoir itself and can be combined with other methods. Therefore, the gas injection displacement technique possesses distinct advantages and holds significant potential for further research. The enhanced gas drainage technology is a fundamental approach to mine gas control. Unlike other gas enhancement methods, such as protective layer pressure relief, hydraulic fracturing, hydraulic punching, water jetting, abrasive jetting, and CO_2-induced phase change fracturing—which alter the stress state or physical properties of the coal reservoir to increase coal seam fractures and permeability, thereby reducing gas flow resistance—the gas injection displacement technique involves injecting "non-methane" gases into the coal body at pressures higher than the gas reservoir pressure. This technique promotes the desorption of adsorbed methane by creating a competitive adsorption effect between the injected gases and the methane adsorbed on the coal matrix surface [11–13], as well as by lowering the methane partial pressure. Additionally, the injected gas, being at a higher pressure than the original reservoir pressure, creates a pressure gradient within the reservoir, driving the gas in the coal seam fracture system from a stagnant state to a flowing state, thus accelerating gas migration within the fractures. The technical principle of this method is fundamentally distinct from other technologies. Its applicability is relatively independent of the physical properties of the coal reservoir itself, and it can be combined with other techniques. As a result, the gas injection displacement method has unique advantages and significant potential for further research (Figs. 1.1 and 1.2).

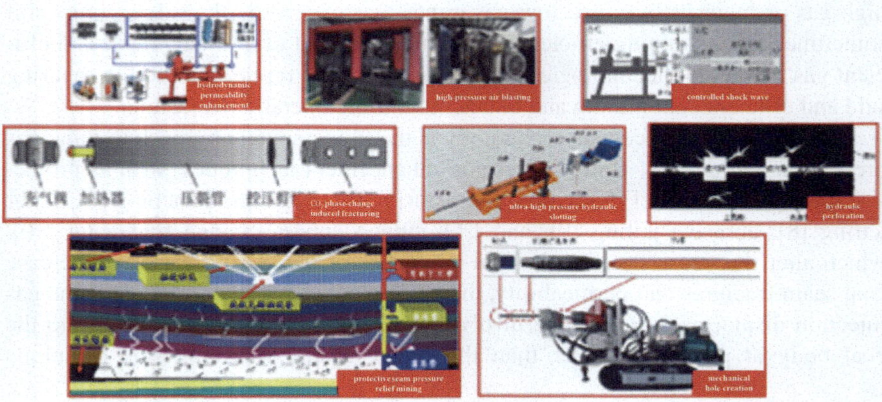

Fig. 1.1 Traditional gas control technology in coal mines

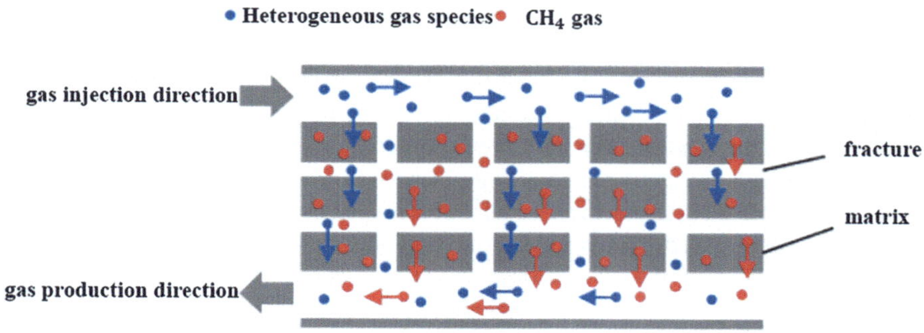

Fig. 1.2 Gas exchange in the coal matrix during gas injection and displacement [14]

In the field implementation of gas injection displacement technology, the "non-methane" gases typically selected are CO_2 and N_2, with some recent studies by Chinese scholars also using air (mainly composed of N_2 and O_2) as the gas source. Worldwide, gas injection displacement has been implemented as a coalbed methane (CBM) production enhancement technique, with engineering trials conducted in various regions, gradually progressing toward commercial application. The CO_2 injection displacement technique (CO_2-ECBM) is the most widely used, with engineering practices primarily focused on increasing coalbed methane production from underground unmineable coal seams [15] and CO_2 geological sequestration [16]. Examples include the micro-experiment at the Allison unit in the San Juan Basin, USA [17], field trials in the Ishikari coalfield in Hokkaido, Japan [18], the third-phase test in the Qinshui Basin, Shanxi, China [19], and the micro-pilot in the Silesian Basin, Poland [20]. However, in nearly all CO_2 injection displacement processes, researchers have found that CO_2 adsorption expansion leads to a reduction in coal body permeability (typically a decrease of 1–2 orders of magnitude) [20–22], and higher CO_2 injection pressures may induce greater volumetric strain, potentially negatively impacting gas production. In CO_2/N_2 mixed gas injection displacement, N_2 plays more of a role as a "permeability restorer." Field test results from Alberta's Fenn Canyon in Canada [23] and the Northern Appalachian Basin in the USA [24] indicate that the N_2 component in the mixed gas significantly improves CBM recovery without affecting the net CO_2 injection rate. Furthermore, field trial results from the Ishikari coalfield in Hokkaido, Japan, suggest that after a period of CO_2 injection, the injection of N_2 can completely reverse the permeability loss caused by CO_2 injection [18]. The N_2 injection displacement technique (N_2-ECBM) has also been applied in certain regions, such as the gas injection tests at the Tiffany unit in the San Juan Basin, USA [25], engineering pilot projects in the Piceance Basin, Colorado, USA [26], and field trials in the Sydney Basin, Australia [27]. Through both field and laboratory studies, scholars have demonstrated that during NN_2 injection for enhanced CBM production, significant improvements in methane recovery have been achieved, and in most coal seams (except for the Piceance Basin pilot test [26]), permeability has been enhanced [22, 28, 29].

Many scholars have conducted extensive research on the evolution of coal permeability during the gas injection displacement process and the competitive adsorption characteristics of mixed gases in coal. However, there is very limited research on the dynamic migration process and patterns of CH_4 gas in coal during the displacement process, and the dynamic migration phenomena of CH_4 gas in coal during injection have not yet been clearly described. Moreover, whether the diffusion behavior of CH_4 gas during the gas injection displacement process can be represented by a single-gas diffusion model, and how it differs from the diffusion of pure CH_4 gas, still requires further investigation. At the same time, there is a lack of corresponding studies on the precise prediction of the gas injection and CH_4 drainage mechanisms and displacement effects in mineable coal seams at the engineering scale.

In summary, gas injection displacement is a coal mine gas enhancement technology with significant engineering application potential, and existing field practices have confirmed its effectiveness. Addressing the aforementioned issues is crucial for a comprehensive understanding of the dynamic migration mechanisms of CH_4 gas in coal during the gas injection displacement process. This book first investigates the dynamic migration patterns of CH_4 gas in coal from both macroscopic and microscopic perspectives. It then explores the field engineering applications of gas injection displacement and the prediction of displacement effects. The anticipated results aim to lay the foundation for the development of a more refined mathematical model for gas injection displacement, enabling a more accurate evaluation of the potential application prospects of this technology under various coal reservoir conditions.

1.2 Research Status

1.2.1 Theoretical Research on Gas Injection Displacement in Coal Seams

In theory, the gas injection displacement process in coal, involving CH_4, is a dynamic process in which a mixed fluid consisting of foreign gases and native CH_4 gas migrates together. For foreign gases, injecting gas into the coal seam at pressures higher than the coal seam gas pressure increases the gas pressure gradient between the injection well/hole and the coal reservoir, causing the foreign gas to flow from the injection boundary into the inter-fractures of the coal reservoir. As injection progresses, the concentration of foreign gases in the inter-fractures gradually increases, and the concentration gradient formed between the inter-fractures and the coal matrix drives the foreign gas to diffuse into and adsorb onto the coal matrix. As for the native CH_4 gas in the coal reservoir, both free and adsorbed CH_4 gases migrate dynamically along the opposite direction of the foreign gas flow and diffusion, eventually being produced at the gas production well/hole.

The fundamental theory of gas injection displacement for coalbed CH_4 can be broadly categorized into three main theories based on the gas mass transfer process and migration behavior: gas adsorption theory [30], Fick's diffusion theory [31], and Darcy's seepage theory [32].

The adsorption phenomenon of gases in coal is primarily explained through theories, such as monolayer adsorption, multilayer adsorption, and micropore filling theory [33–35]. These theories use mathematical models, such as the Langmuir, BET, and DR equations, to fit the adsorption isotherms of gases in coal, thereby revealing the characteristics of gas adsorption behavior in coal. Additionally, through extensive research, scholars have found that gas adsorption in coal can manifest as monolayer adsorption, multilayer adsorption, micropore filling, or even a combination of these forms [36–41]. To provide a more comprehensive description of this process, scholars have developed various composite adsorption models, such as the DA–Langmuir [42], Langmuir–Freundlich [43], and dual Langmuir isothermal adsorption models [44].

In terms of gas diffusion, Fick's law of diffusion states that the diffusion of adsorbates in the adsorbent occurs from areas of high concentration to areas of low concentration, and the diffusion rate is proportional to the concentration gradient. Based on this principle, researchers both domestically and internationally have established various mathematical models to analyze the diffusion behavior of gases in porous media and have proposed several empirical formulas [45–47]. Additionally, they have developed mathematical models to characterize gas diffusion behavior in coal, such as the single-pore diffusion model [48–50], dual-pore diffusion model [51–55], triple-pore diffusion model [56], and dispersive diffusion model [57, 58]. These models help to better understand the diffusion characteristics of gases in coal seams.

Darcy's law assumes that the gas seepage flux is linearly related to the pressure gradient. However, in practice, the flow of gas in coal seams is much more complex. It is not only influenced by the gas pressure gradient in the coal reservoir but also by the stress state of the coal reservoir, the structure of the coal reservoir's pore–fracture system, and the physical properties of the coal reservoir. To more accurately simulate the flow of gas within coal reservoirs, researchers have proposed various modified models, including models that account for the Klinkenberg effect [59], polynomial seepage laws [60], and power-law models describing nonlinear seepage [61]. These models provide more precise descriptions of gas flow in coal seams.

1.2.2 Physical Simulation of Gas Injection Displacement in Coalbed Methane

In previous studies, different scholars have conducted displacement experiments using block coal samples to investigate the overall displacement effect and the evolution of permeability. For example, Liang Weiguo et al. [62, 63] used a self-developed large-scale displacement device to study the effect of CO_2 displacement on coalbed CH_4, concluding that multiple factors collectively determine the displacement efficiency. Wang Liguo [64] varied the injection pressure and stress applied to coal blocks to conduct CO_2/N_2 displacement experiments and compared them with natural discharge experiments. He analyzed the impact of different injection pressures and stress conditions on displacement ratio and breakthrough time for foreign gases, concluding that the CH_4 recovery rate under foreign gas injection conditions was much higher than its single discharge rate. Wu Di [65] studied the influence of volumetric stress, pore pressure, and temperature on the amount of CH_4 displaced. Zhou Junping, Xian Xuefu, and Li Xiaohong [66] researched changes in coal sample permeability during ECBM (enhanced coalbed methane) simulation experiments, finding that coal permeability decreased after CO_2 injection, but when a certain proportion ofNN_2 was mixed with the injected CO_2, the permeability of the coal samples was maintained. Zhang Meihong, Wu Shiyue, et al. [67] analyzed the impact of CO_2 injection on coalbed methane extraction from the perspective of coal's mass transfer coefficient. Connell et al. [68] studied the effects of NN_2 displacement on coal seam gas under various external stress conditions and established a corresponding permeability model for the displacement process. Kumar et al. [69] investigated changes in permeability after injecting CO_2 into coal and found that when injection pressure was below 4 MPa, coal permeability decreased, but when the pressure exceeded 4 MPa, the permeability increased. Wang Haichao et al. [70] studied changes in the coal pore structure and permeability after injecting NN_2, with results indicating a 29.9% improvement in coal permeability.

In addition, many scholars have conducted in-depth studies on the final adsorption states of injected gases and CH_4 after gas injection displacement, which is referred to as the "competitive adsorption" behavior. For example, Zhang Zixu et al. [71] studied the adsorption patterns of CO_2/CH_4 mixed gases under different pressure conditions, finding that CO_2 has a superior adsorption capacity compared to CH_4. Yu Hongguan et al. [72] investigated the isothermal adsorption characteristics of CO_2/CH_4 mixed gases on Jincheng and Lu'an coals, concluding that the adsorption amount of CH_4 in the mixed gas does not increase with pressure. Zhou Junping [73] conducted competitive adsorption experiments with CO_2, CH_4, and N_2 mixed gases, concluding that the adsorption selectivity coefficient is closely related to gas pressure. Yang Hongmin, Wang Zhaofeng, et al. [74] studied the impact of the sequence of gas entry on the final adsorption state through binary gas competitive adsorption and desorption experiments. Their research pointed out that the adsorption and desorption behavior of coal for different gases are not influenced by the order in which gases enter the coal, but rather primarily depend on the state of the gas before and after adsorption. Clarkson and Bustin [75] investigated the competitive adsorption of CH_4 and CO_2 mixed gases in coals of different ranks, with results showing that coal rank has an effect on the selectivity of adsorbed gases. Gruszkiewicz et al. [76] analyzed the adsorption patterns of CO_2/CH_4 mixed gases with equal molar mass, finding that coal has a much higher selective adsorption capacity for CO_2 than for CH_4.

A comprehensive analysis of domestic and international research shows that scholars have conducted in-depth studies on the changes in seepage patterns and competitive adsorption characteristics during the gas injection displacement process, achieving a series of valuable results. However, there is very limited experimental research on the dynamic migration patterns of CH_4 gas and the dynamic production mechanisms of CH_4 gas during the gas injection displacement process.

1.2.3 Study on Gas Diffusion Patterns in Coal

In past research on gas diffusion patterns in coal, studies often focused on the diffusion mechanisms of individual gases. For example, Zhang Dengfeng et al. [77] examined the diffusion process of CH_4 and CO_2 in coals of different ranks, concluding that Fick's law of diffusion could adequately describe the gas diffusion process. They also found that the effective diffusion coefficient exhibits a "U"-shaped relationship with coal rank. Yi Jun [78] and others, after analyzing the micropore structure of coal seams, developed a simplified dual-pore diffusion mathematical model to describe the adsorption and diffusion process of coalbed methane (CBM) in coal micropores. Similarly, Nie Baisheng et al. [79] built a similar model aimed at capturing the adsorption–diffusion behavior of CBM in coal micropores and further estimated the diffusion parameters of gas in coal. Li Zhiqiang et al. [80] conducted comparative experiments on the CH_4 diffusion between coal and anthracite, fitting the experimental data and concluding that the CH_4 diffusion coefficient in coal decreases with increasing diffusion time. Based on this, they established a diffusion model for CH_4 in coal that considers dynamic diffusion coefficients. Liu Mingju and Wei Jianping et al. [81, 82] studied the effects of particle size and temperature on CH_4 diffusion in coal, noting that differences in CH_4 diffusion performance in soft and hard coal particles with different particle sizes are mainly due to differences in the coal particles' pore structure. Additionally, they found that increasing temperature improves the diffusion ability of CH_4 in coal. Charrière et al. [83] studied the impact of pressure on the diffusion rates of CO_2 and CH_4 in coal, concluding that diffusion rates decrease as pressure increases. Pillalamarry et al. [84] analyzed the correlation between a single-pore diffusion model and experimental diffusion results, finding that the model fits better at low-pressure stages. Yang Xin et al. [85] conducted adsorption–desorption diffusion kinetics experiments for CH_4 in coal under different pressures, concluding that the dual-pore diffusion model more accurately describes CH_4 diffusion behavior in coal compared to the single-pore model. Liu Yanwei et al. [56], combining experimental gas diffusion patterns and pore structure characteristics in coal, established a three-layer structural physical-mathematical model based on the continuity diffusion theory. After comparing the model with experimental data, they found that the two matched well. Wang Gongda [55], based on the dual-pore diffusion model for coal, subdivided it into two types and used COMSOL software for numerical simulation, proposing a numerical solution for gas diffusion in coal based on the dual-pore diffusion model.

Overall, most studies indicate that Fick's second law of diffusion can effectively describe the diffusion process of a single gas in coal, leading to widely accepted classical models, such as the "single-pore diffusion model" and the "dual-pore diffusion model." However, for the dynamic diffusion of CH_4 gas during the gas injection displacement process, the following questions are crucial and need to be addressed: Can the dynamic diffusion process of CH_4 gas under gas injection displacement conditions be described using traditional single-gas diffusion laws? Are the diffusion parameters of CH_4 gas under gas injection displacement conditions the same as those during its single-gas diffusion? What are the differences between the dynamic diffusion of CH_4 gas under gas injection displacement conditions and its diffusion as a single gas? These questions require specific experimental research for validation.

1.2.4 Numerical Simulation of CH_4 Displacement in Coal by Gas Injection

Numerical simulation studies of CH_4 displacement in coal seams by gas injection can be broadly categorized into three scales based on the spatial domain of the simulation: microscopic (molecular scale), macroscopic, and engineering-scale simulations.

In the numerical simulation of CH_4 displacement in coal seams at the microscopic scale, Zhang et al. [86] employed the Grand Canonical Monte Carlo (GCMC) method to place CO_2 molecules into pores that had already reached adsorption equilibrium with CH_4. By calculating the final desorbed amount of CH_4 and the adsorbed amount of CO_2 under varying pore pressure conditions, they analyzed the displacement effect of CO_2 on CH_4. The study showed that, under a fixed CO_2 injection pressure, both the amount of CH_4 displaced and the amount of CO_2 sequestered increased with rising porosity. Wu et al. [87] simulated the gas injection process using molecular dynamics (MD) by setting "gas reservoirs" with specific gas pressures on both sides of the pores. They studied the displacement mechanism of heteronuclear gases on CH_4 within carbon nanotubes and concluded that CO_2 reduces CH_4 content via competitive adsorption, whereas N_2 reduces CH_4 content by lowering its partial pressure.

Sun Zhixue et al. [88], using Material Studio, constructed a large anthracite molecular model under specific temperature and pressure conditions and developed a periodic model to investigate the alternating displacement behavior of CO_2 and N_2 on CH_4. The results indicated that when the injection ratio of N_2 to CO_2 was 0.6, the CH_4 desorption rate reached its maximum. Bai et al. [89] developed a coal pore structure model based on coal molecular modeling to analyze CH_4 desorption characteristics after CO_2 and N_2 injection under different temperatures. The results revealed that CO_2 injection was more effective than N_2 in promoting CH_4 desorption. Li Shugang et al. [90] used the GCMC method to construct a CH_4-containing coal adsorption model and investigated the desorption characteristics of CH_4 during gas injection displacement under varying injection pressures, clarifying the intrinsic mechanism by which foreign gases promote the desorption of adsorbed CH_4 molecules in coal.

To characterize the coal macromolecules that constitute the micropores of the coal matrix, researchers commonly employ techniques such as ^{13}C-NMR spectroscopy [91–94], FT-IR [95, 96], and Raman spectroscopy [97–99]. Alternatively, several widely accepted coal macromolecular structures are used, including the Wiser model [100, 101], the van Krevelen model [102], the Given model [103], the Honda model [104], and the Shinn model [105]. To reduce computational load, simplified models such as carbon nanotubes [106] and graphene layers [107] have also been developed. Regarding the pore geometry of the coal matrix, a variety of structural configurations have been proposed, including cylindrical pores [108, 109], slit-shaped pores [110, 111], square pores [112], and sawtooth-shaped pores [113]. In the numerical simulation of CH_4 displacement in coal seams at the nanoscale, extensive research has been conducted, primarily focusing on the competitive adsorption between heteronuclear gases and CH_4 molecules within coal matrix pores. However, the results of these studies often emphasize final adsorption quantities, while overlooking the fact that gas injection displacement at the microscale is a continuous and dynamic process. Under sustained injection, the transport behavior of mixed gases within coal matrix pores should also be dynamic. Therefore, there is a pressing need to construct a more realistic dynamic displacement model to enable accurate microscale simulation of the gas injection displacement process.

For the simulation of CH_4 displacement in coal by gas injection at the macroscopic and engineering scales, most researchers have employed numerical simulation using COMSOL software. Wang Wei et al. [114], under laboratory stress conditions, simulated the displacement of CH_4 in coal pores and fractures by CO_2 based on a constructed permeability model. They conducted numerical simulations at the coal sample scale under different CO_2 injection pressures. The results indicated that with increasing injection pressure, CH_4 production increased significantly during the early stage of CO_2 injection, but in the later stage, CH_4 output tended to decline. Lin et al. [115] used a binary gas transport model to investigate parameters influencing the performance of N_2 injection for CO_2 displacement. The study found that coal seam permeability and CO_2 diffusion coefficient are critical parameters affecting displacement performance, while the effect of the N_2 diffusion coefficient was negligible. Yang Hongmin et al. [116] developed a multi-field coupled mathematical model that comprehensively considers seepage, diffusion, and multicomponent gas adsorption/desorption. They used this model to simulate the dynamic changes in gas emission and extraction flow rates at boreholes during gas injection into coal seams. Wu Jintao et al. [117] proposed a dual-medium mathematical model for simulating gas migration during injection displacement. This model accounts for mixed gas seepage, adsorption/desorption, and diffusion, while also incorporating the sensitivity of coal porosity and permeability. Wang Zhaofeng et al. [118], based on a constructed underground gas injection displacement model, investigated the effective influence radius of gas injection. Wang Gongda, Yang Xin, et al. [59, 60] developed a coupled mathematical model describing CH_4 displacement in coal seams by gas injection. This model was built upon the extended Langmuir model, the P\&M permeability model, and a dynamic Klinkenberg coefficient model, and it comprehensively considered mixed gas adsorption behaviors and the resulting coal swelling effects. Ding Zhiben [119] constructed a mathematical model capable of fully describing the transport behavior of mixed gases in coal. Numerical simulations were conducted under various N_2 injection pressures and initial permeability conditions. By comparing the simulation results with experimental data, the accuracy of the dynamic permeability evolution model was validated.

In summary, numerical simulations of gas injection displacement at the microscopic scale primarily focus on modeling the adsorption and desorption behavior between heteronuclear gas molecules and CH_4 molecules from a molecular perspective. In contrast, simulations at the macroscopic and engineering scales mainly address the seepage behavior of mixed gas flows. However, numerical studies that explicitly consider the dynamic migration behavior of CH_4 during the gas injection displacement process across both scales remain limited.

1.2.5 Field Engineering Applications of Gas Injection for Coalbed Methane Displacement

Globally, gas injection displacement has been implemented as a surface well technology for enhancing coalbed methane (CBM) production, with microscale pilot projects conducted in various locations and gradually advancing toward commercial application. This technology is primarily used to improve the recovery efficiency of CBM from unmineable coal seams and to facilitate geological sequestration of greenhouse gases. At present, documented industrial trials of gas injection displacement within underground coal mines remain relatively limited, with all known implementations occurring in China. In these cases, underground gas injection for CBM is mainly applied in the context of mine gas hazard prevention and control. As for surface well injection projects, countries such as China, the United States, Canada, Japan, and members of the European Union have successively carried out field-scale gas injection displacement trials. In contrast, engineering applications targeting recoverable underground coal seams have thus far only been carried out in China. The specific implementation sites are shown in Fig. 1.3, and the corresponding outcomes are presented in Tables 1.1 and 1.2.

The first gas injection displacement project—known as the Coal-Seq project—was located in the San Juan Basin in New Mexico, USA [120–123], and remains the largest ECBM project to date. In this project, pure CO_2 and pure N_2 were injected into the Allison Unit and Tiffany Unit of the basin, respectively, and CH_4 production was sustained from the beginning of the project through 2001. The results demonstrated that the injection of heteronuclear gases did indeed enhance methane

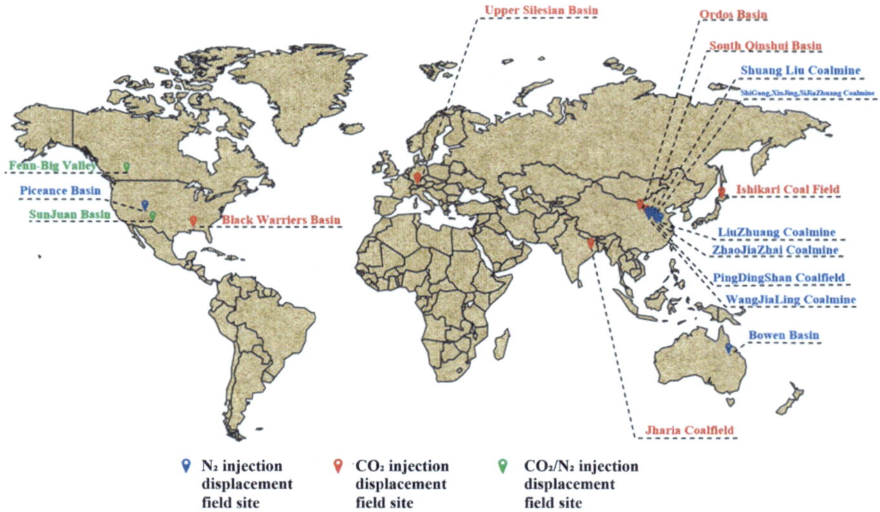

Fig. 1.3 Locations of CO_2/N_2-ECBM project implementation worldwide

Table 1.1 Field test of surface well gas injection for coalbed methane replacement

Place	Project	Year	Inject gas	Well-shaped structure	Injection volume
San Juan Basin, USA	Coal-Seq	1995–2001	N_2	Multiple wells	–
			CO_2	4 Note 16 production	336 kt
Fenn Canyon, Canada	Alberta CO_2/ ECBM	1999	CO_2	Single vertical well	0.19 kt
			13% CO_2/87% N_2		0.11 kt
			53% CO_2/47% N_2		0.12 kt
			N_2		≈0.1 kt
Qinshui Basin, China	Qinshui Basin ECBM	2004	CO_2	Single vertical well	0.19 kt
Bowen Basin, Australia	–	2010	N_2	1 horizontal note 2 production	1.9 kt
Ordos Basin, China	APP ECBM	2011	CO_2	1 horizontal multi-lateral wells	0.46 kt
Ishikari coalfield in Japan, Silesian basin in Poland, Black warrior basin in the United States					

recovery. CO_2 injection led to a reduction in coal seam permeability and injection rate, while N_2 injection resulted in earlier gas breakthrough, which in turn lowered the purity of CH_4 recovery.

Since 1998, under the technical guidance of the Alberta Research Council, a single-well injection experiment was carried out in the Fenn test site in Alberta, Canada [124]. The experiment aimed to achieve dual objectives: reducing

Table 1.2 Field test of downhole gas injection for coalbed methane replacement

Place	Year	Inject gas	Injection time	Injection volume
Yangquan Coal Industry [132]	2010	Air, but only air is considered as N_2 which is not very pure N_2 CO_2	2–8 h	The process measures of various gas injection discharge/pumping have all shown obvious gas stimulation/pumping effect, the efficiency is increased by 2 times, and the cumulative gas extraction volume is increased by 1.95–25.33 times
Fenxi Mining [133]	2017	Continuous and intermittent air injection	16 days	The continuous displacement technology successfully increased the gas extraction capacity by 44%. Although intermittent displacement only increases the extraction capacity by 1%, its displacement efficiency is more significant
Pingdingshan Coal Industry [134]	2006	Air	30 days	Displacement greatly increases the gas concentration and flow rate of the borehole
Xinzheng Coal Power [60]	2018	N_2	33 days	Compared with conventional emissions, N2 displacement significantly increases the net gas content, which is increased by 2–7 times

greenhouse gas emissions and enhancing both the recovery and production of coalbed methane through CO_2 injection. A variety of gases were used during the test, including pure CO_2, flue gas containing N_2, pure N_2, and CO_2-enriched mixed flue gas. The results indicated that injection of pure CO_2 led to a decrease in the absolute permeability of the coal seam, whereas injection of flue gas containing N_2 resulted in an increase in permeability.

In 2001, the European Union launched the Recopol project [125], the first pioneering initiative in Europe dedicated to CO_2 sequestration in coal seams. Conducted in Poland's Upper Silesian Basin, the project aimed to evaluate the technical and economic feasibility of CO_2 storage. As a result of CO_2 injection, CH_4 production increased significantly. From the start of injection in September 2003 until the end of 2004, a total of 760 tonnes of CO_2 were successfully sequestered.

From 2004 to 2005, Japan conducted a field test on CO_2-enhanced coalbed methane recovery in the Ishikari Basin of Hokkaido [18, 126]. The trial was organized by the Ministry of Economy, Trade and Industry and implemented by the General Environmental Technos Co., Ltd. The objective was to develop a cost-effective CO_2 coal seam sequestration technology and to verify its technical and economic feasibility.

Between 2002 and 2006, the Chinese and Canadian governments collaborated on a CO_2-ECBM pilot project in the Qinshui Basin of Shanxi Province, China [127, 128]. This was a single-well injection trial that included stages, such as pre-drainage, CO_2 injection, shut-in, and CBM production. The results demonstrated that CO_2

injection in the southern Qinshui anthracite region not only enhanced coalbed methane recovery but also showed promising potential for CO_2 sequestration.

In addition, field-scale gas injection displacement projects have also been carried out in the Ordos Basin of China [129], the Black Warrior Basin in the United States [130], the Bowen Basin in Australia [27], and the Jharia coalfield in India [131]. These projects will not be discussed in detail here, and the specific information regarding each field trial is presented in Tables 1.1 and 1.2.

1.2.6 Analysis of Existing Problems

Based on the above analysis, it can be concluded that gas injection displacement for coalbed methane (CH_4) is a production enhancement technology that has achieved promising pilot-scale results and holds significant potential for wider application. Previous research has primarily focused on issues such as changes in coal permeability during the displacement process, gas adsorption/desorption behavior, and competitive adsorption among mixed gases, yielding substantial achievements. However, there remain several scientific challenges that urgently need to be addressed:

1. Previous studies on the dynamic migration of CH_4 during the gas injection displacement process have primarily focused on the macroscopic scale. At the microscopic pore scale, research has largely concentrated on the static competitive adsorption behavior of mixed gases, with limited attention given to the dynamic migration of CH_4 molecules. As a result, investigations into the dynamic transport behavior of CH_4 molecules during gas injection displacement from a microscopic pore-scale perspective remain largely unexplored.
2. Unlike the diffusion of a single-component gas, which is driven solely by concentration gradients between the coal matrix interior and exterior, gas diffusion during the displacement process involves bidirectional movement of the injected gas and CH_4 within narrow pore channels. Whether the substantial presence of heteronuclear gases within coal matrix pores influences the dynamic diffusion behavior of CH_4 remains an open question. Moreover, existing theoretical models for gas injection displacement often adopt CH_4 diffusion coefficients derived from single-component gas experiments as direct substitutes for those in displacement scenarios—a practice that is evidently inadequate. Consequently, critical scientific issues such as the applicability of single-component gas diffusion models to CH_4 dynamic diffusion during displacement, as well as the differences between CH_4 diffusion behavior under displacement conditions and its behavior as a single gas, still require further investigation.
3. In the past, gas injection displacement technology has been predominantly applied in surface well operations for enhancing coalbed methane recovery, with target seams typically being unmineable. In recent years, some researchers have extended the application of this technology to the prevention and control of

gas-related hazards in underground mineable coal seams, yielding favorable results. These studies have explored the effects of factors such as injection pressure, injection radius, and injection duration on displacement performance. However, little attention has been paid to how the injection process influences the dynamic migration behavior of CH_4 within coal reservoirs. Moreover, there is a lack of quantitative analysis regarding the variation in displacement efficiency under different operational conditions.

1.3 Research Content and Technical Approach of this Book

1.3.1 Research Content

In response to the aforementioned issues and limitations, this book primarily investigates the multi-scale dynamic migration characteristics of CH_4 in coal during the gas injection displacement process, focusing on the following key aspects:

1. Dynamic Migration Characteristics of CH_4 Molecules During Gas Injection Displacement at the Pore Scale.

   A slit pore displacement model of the coal matrix will be constructed using MS and LAMMPS software to analyze the adsorption, desorption, and displacement behavior of CH_4 molecules within coal matrix slit pores. The study will investigate the dynamic migration patterns of CH_4 molecules under varying conditions, including temperature, slit width, and type of injected gas. In addition, the dynamic migration behavior of heteronuclear gases and the evolution of the displacement front within micropores during the injection process will be examined to further explore the pore-scale dynamic migration characteristics of CH_4 molecules under the influence of gas injection displacement.
2. LBM Simulation of Dynamic CH_4 Migration in Coal During Gas Injection Displacement.

   Based on the lattice Boltzmann method (LBM), an NSE-ADE coupled LB model will be developed to account for the directional flow of mixed gases and the mass transfer between the coal matrix and its surroundings during gas injection displacement. This model will be used to simulate and investigate the dynamic migration behavior of CH_4 gas throughout the injection process. The effects of factors such as injection flow rate and type of injected gas will be examined in detail. Building upon the above results, a comprehensive analysis will be conducted on the dynamic migration and production mechanisms of CH_4 in coal during gas injection displacement.
3. Experimental Study on the Dynamic Migration Behavior of CH_4 in Coal Particles During Gas Injection Displacement.

   Expanding the research perspective to the macroscopic level, a self-developed experimental system for measuring the dynamic migration characteristics of CH_4 in coal particles during gas injection displacement will be employed to quantitatively investigate the CH_4 migration behavior under various conditions. Based on the macroscopic experimental results, the applicability of different single-gas diffusion models in characterizing the dynamic diffusion behavior of CH_4 during the displacement process will be theoretically evaluated. Additionally, the effects of injection flow rate and type of injected gas on the variation of diffusion parameters in each model will be examined. Furthermore, by comparing the experimental results of single-gas CH_4 diffusion and CH_4 dynamic diffusion during displacement, the differences between the two diffusion behaviors will be analyzed.

4. Engineering Application and Displacement Performance Prediction of Underground N_2 Injection for CH_4 Displacement in Coal Seams.

   Using a self-designed underground gas injection displacement system based on a "pressurization–gas storage–injection" process, this study investigates the engineering characteristics of CH_4 dynamic migration in coal reservoirs under N_2 injection conditions. A quantitative analysis is conducted to compare the CH_4 enhancement effects of "gas injection displacement" versus "conventional emission" processes, as well as "intermittent injection" versus "continuous injection" methods. The impact of injection flow rate on CH_4 enhancement performance is also evaluated. Based on the engineering application results, the underlying mechanism of CH_4 enhancement via N_2 injection at the engineering scale is explored. Additionally, a preliminary investigation is carried out into the application of deep learning algorithms for predicting gas injection displacement performance at the engineering scale, along with optimization of the algorithmic model.

1.3.2 Technical Approach

This book adopts an integrated approach combining numerical simulation, experimental investigation, theoretical analysis, and field application to study the dynamic migration characteristics of CH_4 gas in coal during the gas injection displacement process. First, gas injection displacement models were developed at multiple scales, including a microscopic slit pore model capable of providing a continuous pressure gradient, and an NSE-ADE coupled Lattice Boltzmann (LB) model that considers multicomponent gas flow and mass transfer between coal matrix and pore spaces. Based on these models, numerical simulations were conducted to analyze the CH_4 gas migration process during gas injection displacement at different scales, thereby revealing the dynamic migration behavior of CH_4 under various conditions. Second, a self-developed experimental system for measuring the dynamic migration of CH_4 gas in coal particles during gas injection displacement was used to perform macroscopic scale experiments. These experiments helped clarify the CH_4 migration patterns in coal under different influencing factors. Furthermore, a CH_4 gas diffusion model was introduced that incorporates gas diffusion–relaxation characteristics, enabling the exploration of intrinsic feedback mechanisms of CH_4 dynamic

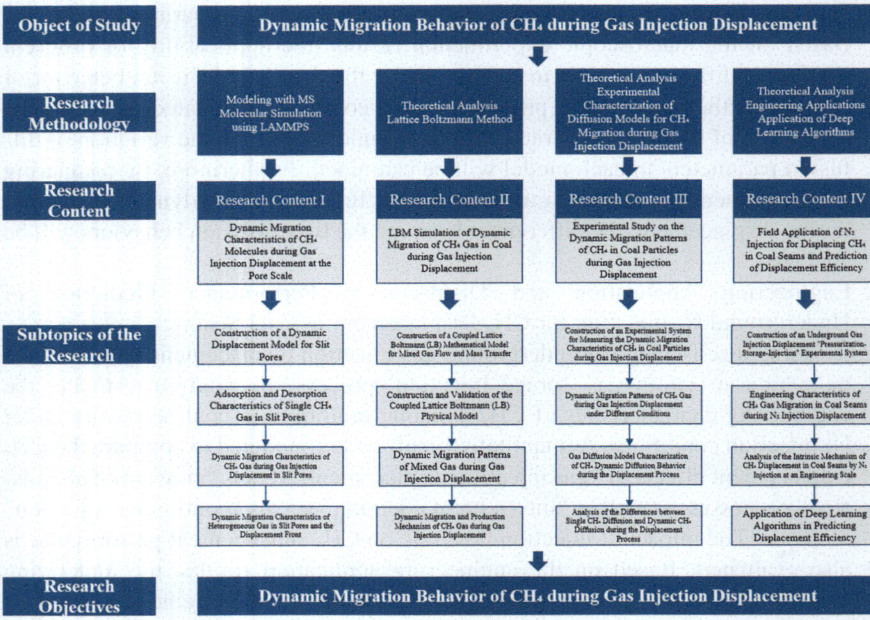

Fig. 1.4 Technical route

diffusion during displacement. Finally, a field engineering application was carried out to assess the effectiveness of N_2 injection in promoting CH_4 extraction from coal seams. The engineering-scale characteristics of CH_4 gas migration under N_2 injection conditions were investigated. Additionally, deep learning algorithms were employed to predict the outcomes of field-scale gas injection displacement, achieving effective forecasting of performance. The technical approach of this study is illustrated in Fig. 1.4.

Chapter 2
Dynamic Migration Characteristics of CH_4 Molecules During Gas Injection Displacement at the Pore Scale

2.1 Introduction

Coal is a complex porous medium with a highly anisotropic internal pore structure, shaped by factors, such as burial depth and formation processes. The pore size distribution within coal spans a wide range—from the nanometer to the micrometer scale. The pores in the coal matrix provide an extensive specific surface area, offering numerous adsorption sites for adsorbed gas. To simplify research problems, many scholars have proposed various classification schemes based on pore size, as shown in Table 2.1. Studies have demonstrated [135] that coal contains a large number of micropores and small pores, some mesopores, and a few macropores. Among these, micropores and small pores serve as the primary regions for gas adsorption, mesopores provide the main pathways for gas diffusion, and macropores function as the dominant channels for gas seepage. Together, micropores and mesopores form the principal storage space for gases in coal.

The nanoscale pores within the coal matrix serve as the primary sites for CH_4 storage and diffusion behavior [148]. At the pore scale, the entire gas injection displacement process in coal can be viewed as the interaction between the injected mixed gas molecules and CH_4 molecules within the micropores of the coal matrix. Conventional macroscopic experimental methods are incapable of directly observing the displacement characteristics and mechanisms occurring inside these pores. In contrast, molecular simulation techniques provide an effective approach to study the gas transport behavior during the injection-driven displacement process at the microscopic pore level. Most researchers simulate pore-scale gas displacement behavior using two primary methods:

(a) The Grand Canonical Monte Carlo (GCMC) method is employed to directly introduce heteronuclear gas molecules into pores that have already reached adsorption equilibrium with CH_4 molecules. By calculating the amount of CH_4

G. Wang, H. Gong, *Dynamic Migration Characteristics of CH4 During Gas Injection Displacement in Coal Seams*,
https://doi.org/10.1007/978-981-95-8879-4_2

Table 2.1 Coal pore classification statistics table [136–147]

Categorization	Micropores/nm	Small holes/nm	Medium holes/nm	Large holes/nm
Hodot (1961)	<10	10–100	100–1000	>1000
Dubinin (1963)	<2	2–20	–	>20
Gan H (1972)	<1.2	1.2–30	–	30–2960
IUPAC (1978)	<2	–	2–50	>50
Chandra G(1987)	<0.8	0.8–2	2–50	>50
Zhu Zhipei (1982)	<12	12–30	–	>30
Fu Shunyuan (1985)	<8	8–100	–	>100
Yang Simin (1991)	<10	10–50	50–750	>1000
Wu Jun (1991)	<10	10–100	100–1000	1000–15,000
Yu Qixiang (1992)	<10	10–100	100–1000	1000–100,000
Qin Yong (1994)	<10	10–50	50–450	>450
Clarkson C (1999)	<2	–	2–50	–

desorbed and the amount of heteronuclear gas adsorbed within the pores, researchers can analyze the displacement characteristics of heteronuclear gases on CH_4 molecules, as illustrated in Fig. 2.1a.

(b) Gas reservoirs are connected to one or both sides of the pore structure, with a certain amount of CH_4 molecules pre-adsorbed inside the pore. A fixed number of heteronuclear gas molecules are placed in the gas reservoirs to represent the injection pressure, thereby simulating the displacement behavior at the pore scale. This method is essentially an improvement upon the first approach, as illustrated in Fig. 2.1b.

However, the displacement behavior of heteronuclear gases toward CH_4 within coal pores is a dynamic process governed jointly by the pressure gradient and competitive adsorption. The transport of CH_4 molecules under displacement conditions is likewise a dynamic phenomenon—yet this crucial aspect has been largely overlooked in previous studies. Therefore, there is an urgent need for a pore-scale displacement model that more accurately reflects the actual injection-driven process occurring in coal matrix pores. In this chapter, we address the limitations of previous research by constructing a coal matrix slit pore displacement model using MS and LAMMPS software, which allows for the application of a continuous pressure gradient. Based on this model, we investigate the adsorption, desorption, and dynamic transport behavior of CH_4 molecules within the coal matrix slit pore under displacement conditions, thereby exploring the underlying mechanism of CH_4 production at the pore scale during gas injection displacement.

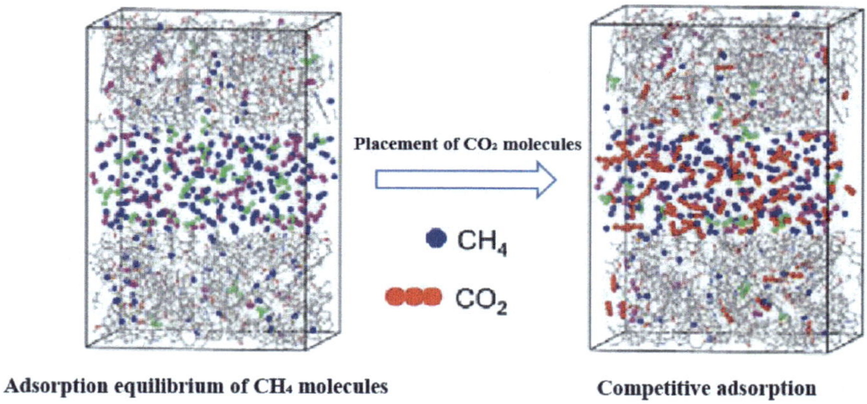

(a) model①[149][150]

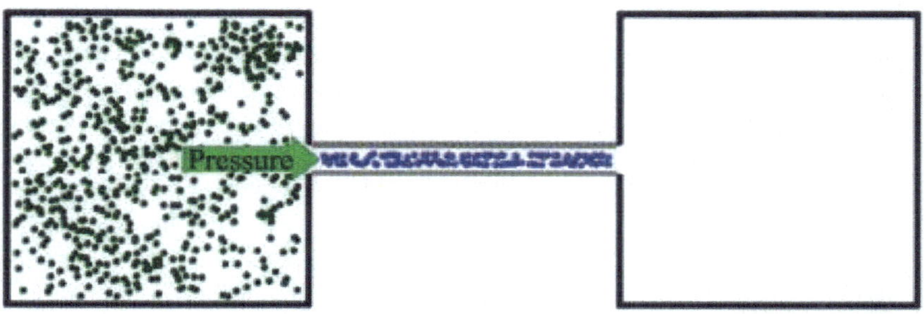

(b) model②[87][150]

Fig. 2.1 Molecular models of displacement constructed in previous studies. (**a**) Model ① [149]. (**b**) Model ② [87, 150]

2.2 Fundamental Principles of Molecular Simulation in Gas Injection Displacement Processes

2.2.1 Molecular Force Field

In molecular simulations, a molecular force field is a mathematical model used to describe the interactions within and between molecules, enabling the simulation of molecular structure and dynamic behavior. The completeness and accuracy of a force field largely determine the reliability of the simulation results. Force fields are generally classified based on whether they account for the formation or breaking of chemical bonds. Reactive force fields are designed for systems where chemical bonds may form or break during the simulation. In contrast, classical force fields are used for stable molecular systems in which chemical bonds remain unchanged. Since this chapter does not involve chemical reactions, only classical force fields are employed to describe the interactions between atoms.

Typically, the internal energy of a molecule arises from several types of intramolecular potential energy, including bond stretching energy (E_{bond}), angle bending energy (E_{angle}), out-of-plane distortion energy (E_{impro}), and dihedral torsion energy (E_{torsion}). As for intermolecular interactions, they primarily involve van der Waals energy (E_{vdw}) and Coulombic electrostatic energy (E_{coul}). These energy components collectively determine the structural stability and interaction characteristics of the molecule. The general form of a force field can be expressed as Eq. (2.1):

$$E_{total} = E_{angle} + E_{bond} + E_{impro} + E_{torsion} + E_{vdw} + E_{coul} \tag{2.1}$$

In this equation, E_{total} represents the total potential energy.

2.2.1.1 Bond Angle Bending Energy (E_{angle})

Bond angle bending energy (E_{angle}) is an energy term in the molecular force field used to describe how the angular relationships between adjacent atoms contribute to the total potential energy of a molecule. Specifically, E_{angle} quantifies the energy change caused by deviations of bond angles from their ideal values. This concept addresses the angles formed between atoms that are connected by chemical bonds within a molecule. When these angles deviate from their equilibrium or ideal values, E_{angle} captures the resulting impact on the molecule's potential energy. This energy is typically calculated using an angular potential function, most commonly modeled by a harmonic oscillator, as shown in Eq. (2.2).

$$E_{angle} = \sum_{angles} \frac{K_\theta}{2} (\theta_i - \theta_{i0})^2 \tag{2.2}$$

In this equation, K_θ is the bond angle force constant, representing the stiffness of the angle between adjacent atoms. θ_i and θ_{i0} denote the actual bond angle and the corresponding ideal bond angle for the i angle, respectively. This function describes the potential energy of a harmonic oscillator, quantifying the energy variation resulting from deviations of the actual angle from the ideal angle. In molecular simulations, this type of potential energy function is used to calculate the total potential energy of the molecule, thereby influencing its conformation and dynamic behavior.

2.2.1.2 Bond Stretching Energy (E_{bond})

The bond stretching energy (E_{bond}) is used to describe the effect of bond length variations on the total potential energy of a molecule. Specifically, E_{bond} represents the energy change caused by the deviation between the actual bond length and the ideal bond length. When the actual bond length deviates from the ideal value, E_{bond} quantifies the resulting influence on the molecular potential energy. In classical force fields, this energy term is typically calculated using the harmonic oscillator model, as shown in Eq. (2.3).

$$E_{bond} = \sum_{bonds} K_b \left(b_i - b_{i0}\right)^2 \tag{2.3}$$

In this equation, K_b is the bond force constant, representing the stiffness of the bond between adjacent atoms; and b_i and b_{i0} denote the actual and the ideal bond lengths of the i bond, respectively.

2.2.1.3 Out-of-Plane Bending Energy (E_{impro})

Out-of-plane bending energy (E_{impro}) describes the energy associated with the deviation of an atom from a defined molecular plane. This term is particularly important in systems where planar configurations are preferred, such as in aromatic rings or conjugated molecules. When an atom bonded to a planar structure deviates from the plane formed by its neighboring atoms, E_{impro} quantifies the resulting increase in potential energy. It is typically modeled using a harmonic potential function, similar to angle and bond stretching terms, as shown in Eq. (2.4).

$$E_{impro} = \frac{1}{2}\sum_{i} K_{\chi} \chi^2 \tag{2.4}$$

In this equation, K_{χ} is the force constant for the out-of-plane bending term, representing the stiffness of the deviation, and χ is the angle of out-of-plane bending.

2.2.1.4 Torsional Energy ($E_{torsion}$)

Torsional energy ($E_{torsion}$) is used to describe the effect of dihedral angle torsion between four adjacent atoms (usually four nonlinearly connected atoms) on the total potential energy of a molecule. The dihedral angles between atoms can significantly influence the overall molecular conformation; therefore, $E_{torsion}$ quantifies the contribution of such conformational changes to the molecule's total potential energy. The inclusion of this energy term is essential for accurately capturing and simulating intramolecular conformational variations in molecular simulations. It is typically expressed as shown in Eq. (2.5).

$$E_{torsion} = \frac{1}{2}\sum_{n} V_n \left(1 + \cos\left(n\theta - \delta\right)\right) \tag{2.5}$$

In this equation, V_n represents the torsional barrier height, indicating the maximum energy barrier encountered during the torsion process. It influences the resistance to torsion—higher barrier values signify greater resistance, making torsional motion more difficult. n denotes the periodicity of the torsion, reflecting how frequently the torsional angle repeats between adjacent atoms; it is typically taken as $n = 3$. θ is the torsional angle, referring to the rotational angle between the planes formed by the three bonds connecting the four atoms. δ is the phase angle, which determines the starting position of the potential energy function in space. Adjusting the phase angle shifts the phase of the energy function and affects the relative positioning of the torsional energy.

2.2.1.5 Van der Waals Interaction Energy (E_{vdw})

Van der Waals interaction is a combination of attractive and repulsive forces that typically occurs between nonpolar molecules or atoms. This interaction arises from the motion of electron clouds, which leads to a temporary uneven distribution of electrons in space, inducing polarization in neighboring molecules. This transient charge distribution generates both attractive and repulsive forces, forming what is known as the Van der Waals force. The Van der Waals interaction energy (E_{vdw}) represents this effect and is also referred to as non-bonded potential energy.

A commonly used potential function to describe Van der Waals interactions is the Lennard-Jones potential function [151]. This function consists of two main components: an attractive term (r^{-6}) and a repulsive term (r^{-12}), as shown in Eq. (2.6).

$$E_{vdw} = 4\varepsilon_{ij}\left[\left(\frac{\sigma_{ij}}{r_{ij}}\right)^{12} - \left(\frac{\sigma_{ij}}{r_{ij}}\right)^{6}\right] \tag{2.6}$$

In the above equation, r represents the distance between two interacting nonpolar atoms or molecules; ε is the depth of the potential well, indicating the strength of the interaction; and σ is the distance at which the potential energy is zero.

This function exhibits a characteristic Van der Waals potential well, where attractive forces dominate at shorter distances and repulsive forces dominate at greater distances. It effectively models the two main components of Van der Waals forces: attraction and repulsion. Van der Waals interactions play a critical role in molecular interactions, significantly influencing molecular structure, stability, and physical properties. In molecular force fields, Van der Waals interaction energy is a key component for simulating and describing non-covalent interactions between molecules.

In this book, the Lorentz–Berthelot mixing rules [152] are used to determine the potential parameters between different types of particles, as expressed in Eqs. (2.7) and (2.8).

$$\varepsilon_{ij} = \sqrt{\varepsilon_{ii}\varepsilon_{jj}} \tag{2.7}$$

$$\sigma_{ij} = \frac{\sigma_{ii} + \sigma_{jj}}{2} \tag{2.8}$$

2.2.1.6 Coulombic Interaction Energy (E_{coul})

Coulombic interaction energy (E_{coul}) is an energy term in molecular force fields used to describe the interactions between charged particles. It accounts for the electrostatic interactions between charges—repulsive forces occur between particles with like charges, while attractive forces occur between particles with opposite charges. The expression for Coulombic interaction energy is based on Coulomb's law, which describes the interaction between two electric charges, and can be expressed as Eq. (2.9):

$$E_{coul} = \sum_{i=1}^{N}\sum_{j=i+1}^{N} \frac{1}{4\pi\varepsilon_0}\frac{q_i q_j}{r_{ij}} \tag{2.9}$$

In the above equation, ε_0 represents the vacuum permittivity, with a value of 8.854×10^{-12} Fm^{-1}; q_i and q_j are the charges of the two interacting particles; and r_{ij} is the distance between the two charged particles.

In molecular force fields, Coulombic interaction energy is typically used in conjunction with other energy terms (such as van der Waals interaction energy) to provide a comprehensive description of both intra and intermolecular interactions.

2.2.2 *Molecular Dynamics Simulation*

The molecular dynamics simulation method assumes that all particles (molecules or atoms) in the system move under the influence of force fields according to classical laws of motion, namely Newton's laws. The core idea of this method is to solve the Newtonian equations of motion in the constructed force field system, obtain the motion states of particles in the system, and calculate the motion states of some or all particles at future times using a reasonable time step. Through this process, statistical methods are used to compute the required physical quantities, thereby studying the microscopic characteristics of the system. The general procedure of molecular dynamics simulation can be summarized as shown in Fig. 2.2.

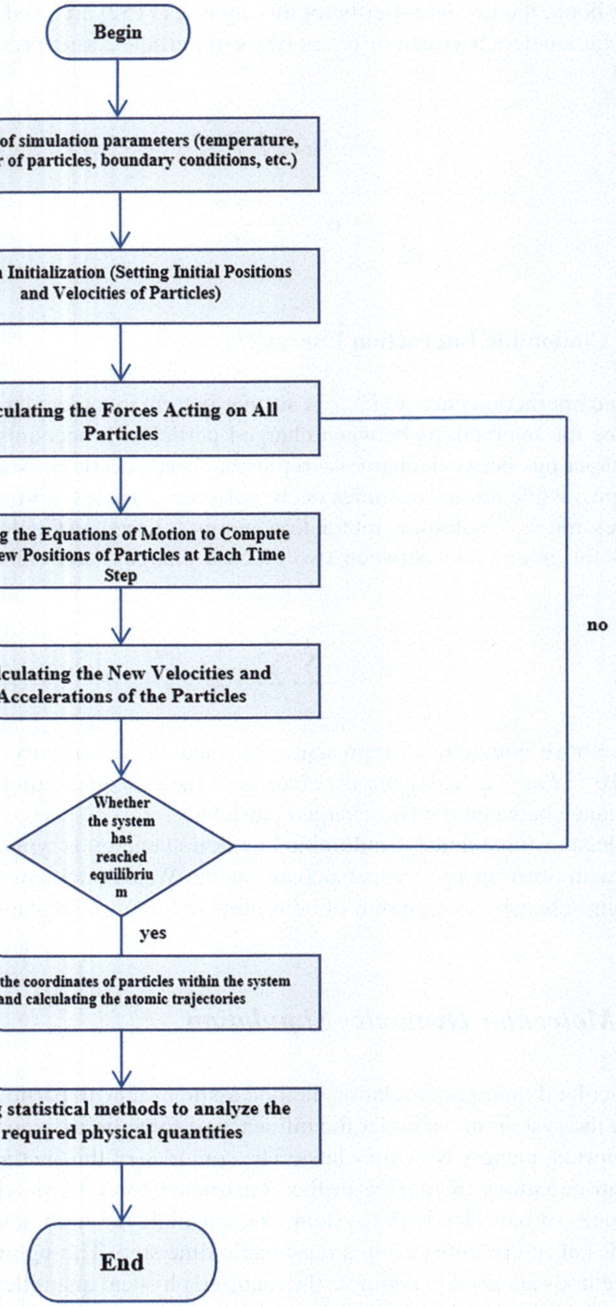

Fig. 2.2 Calculation flow of molecular dynamics simulation

2.2.2.1 Statistical Ensemble

The results obtained from macroscopic experiments represent the averaged properties of the measured values, meaning they reflect the average behavior of a large number of particles. If computer simulations are to serve as a counterpart to macroscopic experiments, the measured properties of individual particles must also be averaged accordingly. However, in molecular dynamics simulations, directly simulating the motion of every particle in a large-scale system is typically very complex and computationally demanding. Therefore, the concept of an ensemble is introduced. All microscopic systems that constitute an ensemble are defined to have the same number of molecules, temperature, and pressure as the real system being represented. This allows researchers to focus only on a representative group of systems that share the same macroscopic properties, thus facilitating the study of their average behavior and properties without the need to track the exact state of each particle. At the macroscopic scale, systems in equilibrium typically consist of multiple microstates. The ensemble average of a certain dynamic property A can be calculated using the following equation (Eq. 2.10):

$$\langle A \rangle = \sum_i A_i p_i \tag{2.10}$$

In this equation, A_i represents the value of property A in the *ith* microstate, and p_i is the probability of occurrence of the *ith* microstate. The notation $<A>$ denotes the ensemble average.

The classification of ensembles is primarily based on whether the number of particles in the system is variable. They can be broadly divided into two main categories. The first category is closed systems, which include the microcanonical ensemble (NVE), the canonical ensemble (NVT), and the isothermal-isobaric ensemble (NPT). In these ensembles, the number of particles remains constant. The second category is open systems, with the grand canonical ensemble (μVT) being the most commonly used. This ensemble allows the number of particles to fluctuate under specified conditions. These ensembles provide the theoretical framework for studying the thermodynamic and statistical properties of systems under various conditions:

Microcanonical Ensemble: In the microcanonical ensemble, the system does not exchange energy or particles with its surroundings, and its volume remains fixed. This ensemble is suitable for describing a system that is completely isolated from the external environment, with both energy and particle number conserved.

Canonical Ensemble: In the canonical ensemble, the system exchanges energy with a heat reservoir while maintaining constant volume and particle number. This ensemble is typically used to describe a system that can exchange energy with its surroundings, but whose volume and number of particles remain unchanged.

Isothermal–Isobaric Ensemble: In the isothermal–isobaric ensemble, the system can exchange energy with both a heat reservoir and a pressure reservoir, allowing the system's volume to fluctuate. This ensemble is suitable for describing systems under constant temperature and pressure conditions.

Grand Canonical Ensemble: In the grand canonical ensemble, the system exchanges energy with a heat reservoir and particles with a particle reservoir, and its volume may also vary. This ensemble is commonly used to describe systems where particle exchange occurs, with the chemical potential remaining constant.

This study primarily focuses on the dynamic transport behavior of CH_4 gas within coal matrix pores during the displacement process. The entire system is required to maintain constant temperature and pressure conditions. Specifically, the isothermal–isobaric ensemble (NPT) is employed during the construction of the coal macromolecular layers, while the canonical ensemble (NVT) is used to investigate the dynamic transport behavior of CH_4 gas during the displacement process.

2.2.2.2 Equations of Motion

In molecular dynamics simulations, it is assumed that the motion of particles can be treated using classical mechanics. For an isolated system containing a total of N particles, the total energy of the system is composed of the sum of the total potential energy and total kinetic energy of all particles. This is typically represented by the Hamiltonian, as shown in Eq. (2.11).

$$H = \sum_{i=1}^{N} \frac{p_i^2}{2m_i} + U\left(q_1, q_2, \ldots, q_N\right) \tag{2.11}$$

In this equation, p_i represents the momentum of the ith particle; m_i is the mass of the ith particle; and $U(q_1,q_2,...,q_N)$ denotes the total potential energy of the system (denoted as U_{total}, which is a function of the particle positions). Here, $q_1,q_2,...,q_N$ are the spatial coordinates of the particles in the system, meaning that the potential energy depends on the specific positions of each particle. This relationship is expressed in Eq. (2.12).

$$U_{total} = \sum_{i=1}^{N-1} \sum_{j=i+1}^{N} u_{ij}\left(r_{ij}\right) \tag{2.12}$$

Based on classical mechanics theory, the force acting on any particle i in the system is given by the negative gradient of the potential energy:

$$F_i = -\nabla U_{total} = -\frac{\partial U_{total}}{\partial r_i} = -\left(\vec{i}\frac{\partial}{\partial x_i} + \vec{j}\frac{\partial}{\partial y_i} + \vec{k}\frac{\partial}{\partial z_i}\right) U_{total} \tag{2.13}$$

According to Newton's second law, the acceleration of particle i is given by:

$$\vec{a}_i = \frac{\vec{F}_i}{m_i} \tag{2.14}$$

By integrating over time, the velocity and position of particle i after a time interval Δt are given by:

$$\frac{d^2}{dt^2}\vec{r}_i = \frac{d^2}{dt^2}\vec{v}_i = \vec{a}_i \tag{2.15}$$

$$\vec{v}_i = \vec{v}_{0i} + \vec{a}_i t \tag{2.16}$$

$$\vec{r}_i = \vec{r}_{0i} + \vec{v}_{0i} + \frac{1}{2}\vec{a}_i t^2 \tag{2.17}$$

In Eqs. (2.12)–(2.17), U_{total} denotes the total potential energy of the system; $\vec{F}_i$, $\vec{a}_i$, $\vec{r}_i$, and $\vec{v}_i$ represent the force, acceleration, position, and velocity of particle i, respectively; and $\vec{r}_{0i}$ and $\vec{v}_{0i}$ denote the initial position and initial velocity of particle i, respectively.

In molecular dynamics simulations, to obtain the trajectory information of particle i at various time points, it is necessary to solve the equations of motion. Currently, the mainstream solution algorithms include the Verlet algorithm, the leap-frog method, and the velocity Verlet algorithm.

Verlet Algorithm

The Verlet algorithm [153], proposed in 1967, is one of the most widely used methods due to its advantages of low memory usage and ease of implementation. By performing a Taylor series expansion of the particle's position, the positions of a particle at times $t - \Delta t$ and $t + \Delta t$ can be expressed in terms of its position at time t, as shown in Eqs. (2.19) and (2.20).

$$\vec{r}(t+\Delta t) = \vec{r}(t) + \frac{d}{dt}\vec{r}(t)\Delta t + \frac{1}{2!}\frac{d^2}{dt^2}\vec{r}(t)(\Delta t)^2 + \cdots \tag{2.18}$$

$$\vec{r}(t-\Delta t) = \vec{r}(t) - \frac{d}{dt}\vec{r}(t)\Delta t + \frac{1}{2!}\frac{d^2}{dt^2}\vec{r}(t)(\Delta t)^2 + \cdots \tag{2.19}$$

By adding the two equations and rearranging the terms, the position of the particle at time $t + \Delta t$ can be predicted based on its positions at times t and $t - \Delta t$, as shown in Eq. (2.20).

$$\vec{r}(t+\Delta t)=2\vec{r}(t)-\vec{r}(t-\Delta t)+\frac{d^2}{dt^2}\vec{r}(t)(\Delta t)^2+\cdots \tag{2.20}$$

By subtracting the two equations and rearranging the terms, the velocity of the particle at time t can be calculated based on its positions at times $t-\Delta t$ and $t+\Delta t$, as shown in Eq. (2.21).

$$v(t)=\frac{dr}{dt}=\frac{1}{2\Delta t}\left[r(t+\Delta t)-r(t-\Delta t)\right] \tag{2.21}$$

It should be noted that the velocity of a particle can only be calculated after its position has been determined. Moreover, in most cases, the value of Δt is very small, which can easily lead to computational errors.

Leap-Frog Method

The leap-frog method is an improved algorithm developed by Hockney in 1970 [154] based on the Verlet algorithm. This method eliminates the need for interpolation and simplifies Eq. (2.18) to a second-order form, as shown in Eq. (2.22).

$$r(t+\Delta t)=r(t)+\frac{d}{dt}r(t)\Delta t+\frac{1}{2!}\frac{d^2}{dt^2}r(t)(\Delta t)^2 \tag{2.22}$$

$$r(t+\Delta t)=r(t)+\Delta t\left(v(t)+\frac{a(t)}{2}(\Delta t)\right) \tag{2.23}$$

According to Newton's equation of motion, it can be expressed as:

$$v\left(t+\frac{1}{2}\Delta t\right)=v(t)+a(t)\frac{\Delta t}{2} \tag{2.24}$$

$$v\left(t-\frac{1}{2}\Delta t\right)=v(t)-a(t)\frac{\Delta t}{2} \tag{2.25}$$

Substituting Eq. (2.24) into Eq. (2.23), the expression for the position of a particle at any time is:

$$r(t+\Delta t)=r(t)+v\left(t+\frac{\Delta t}{2}\right)\Delta t \tag{2.26}$$

Subtracting Eq. (2.25) from Eq. (2.24) yields the expression for the velocity of a particle at any given time:

$$v\left(t+\frac{1}{2}\Delta t\right)=v\left(t-\frac{1}{2}\Delta t\right)+a(t)\Delta t \tag{2.27}$$

Velocity Verlet Algorithm

The velocity Verlet algorithm [155], developed by Swope in 1982, is an advancement of the original Verlet algorithm. It was specifically designed to address the shortcomings of both the Verlet and leap-frog methods. This algorithm improves the simulation by simultaneously updating and storing the velocity, acceleration, and position of particles at adjacent time steps, making it more suitable for molecular dynamics simulations that require accurate velocity information for computing temperature, kinetic energy, and transport properties.

The position of a particle at any given time within the system is expressed by Eq. (2.28):

$$r(t+\Delta t)=r(t)+\Delta t\left(v(t)+\frac{a(t)}{2}(\Delta t)\right) \tag{2.28}$$

The velocity of a particle at any given time within the system is expressed by Eq. (2.29):

$$v(t+\Delta t)=v(t)+\frac{a(t)}{2}\Delta t+\frac{a(t+\Delta t)}{2}\Delta t \tag{2.29}$$

This book adopts the velocity Verlet algorithm to solve Newton's equations of motion.

2.2.2.3 Boundary Conditions

In molecular dynamics simulations, boundary conditions are a method for handling the edges of the simulated system in order to mimic a larger-scale system and avoid artifacts introduced by the system boundaries. These conditions define the behavior of particles when they approach or cross the boundaries of the simulation box. One of the most commonly used boundary conditions is the periodic boundary condition (PBC) [156].

Periodic boundary conditions assume that the system is periodic, meaning the boundaries of the system extend infinitely in one or more directions. When a particle crosses the boundary of the system, it reappears on the opposite side, thereby simulating an infinitely large periodic system. Taking two-dimensional periodic boundary conditions as an example, as shown in Fig. 2.3, the macroscopic system can be considered as an assembly of basic units of molecular systems with identical

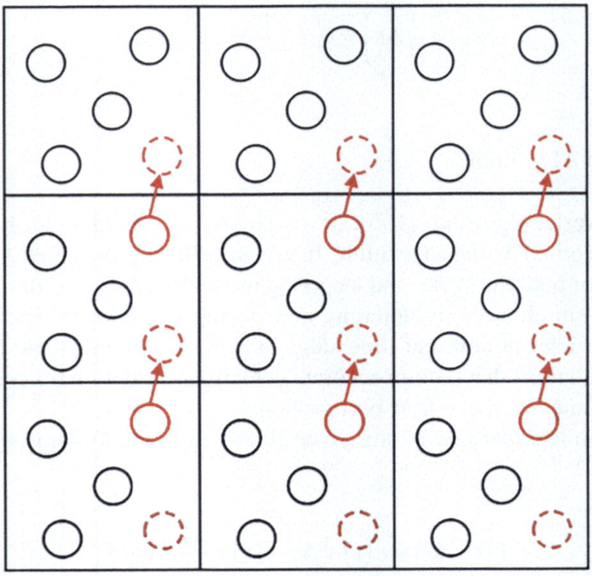

Fig. 2.3 Two-dimensional periodic boundary conditions

properties. These basic units are called unit cells, as represented by the small square in Fig. 2.3. During molecular dynamics simulations, when any particle in a unit cell moves to a neighboring unit cell, an identical particle will simultaneously enter the original unit cell at the corresponding position, thus maintaining the stability of the unit cell's state. Therefore, the introduction of periodic boundary conditions allows the simulation of a small portion of a system to yield the macroscopic properties of the actual system. However, molecular dynamics simulations do not always require periodic boundary conditions. In this study, due to the directional flow behavior of gas induced by gas injection displacement within coal matrix pores, non-periodic boundary conditions are applied in the direction of gas displacement (i.e., the *x*-direction).

2.2.3 Monte Carlo Simulation

Monte Carlo simulation is a numerical technique based on probability and statistics, which solves complex problems through random sampling and statistical analysis. Its core concept involves using a random number generator to perform a large number of random samples in order to approximate the solution of a given problem. In molecular simulation, the purpose of employing the Monte Carlo method is to approximate the properties of a system through random sampling, to reveal its statistical behavior, explore the energy landscape, and simulate the dynamic evolution under non-equilibrium conditions. This provides researchers with a powerful tool for understanding and modeling the complexity of molecular systems.

2.2.3.1 Metropolis Sampling Method

The Metropolis method is a specific algorithm within the Monte Carlo simulation framework and is widely used for simulating the equilibrium behavior of statistical mechanical systems. Specifically, the Metropolis method is a type of Markov Chain Monte Carlo (MCMC) algorithm, whose core idea is to define an acceptance probability to control the transition between states, thereby performing random sampling in the system's state space. The form of this acceptance probability is usually given by the Metropolis criterion [157], which determines whether a proposed state is accepted as the next state.

The basic steps of the Metropolis sampling method are as follows:

(a) Initial State: Select the initial state of the system.
(b) Propose a New State: Propose a new state based on certain rules, such as by altering specific parameters within the system.
(c) Calculate Acceptance Probability: Calculate the probability of accepting the proposed state based on the Metropolis criterion.
(d) Determine State Transition: Use a random number and the acceptance probability to decide whether to accept the proposed state. If accepted, transition to the new state; otherwise, retain the current state.
(e) Repeat the Process: Repeat the above steps to generate a sequence of states.

The Metropolis method is primarily applied in Monte Carlo simulations within statistical mechanics, especially for modeling the equilibrium behavior of systems in molecular simulations. It is used to simulate the properties of various phases of matter—gases, liquids, and solids—including temperature, pressure, heat capacity, and more. In summary, the Metropolis method is a critical implementation of the Monte Carlo simulation approach. By performing random sampling in the state space, it allows the system to transition between different states with a certain probability, thereby effectively modeling the system's equilibrium behavior.

2.2.3.2 Monte Carlo Simulation Calculations

According to practical requirements, the Monte Carlo simulation method can be applied to calculate different ensembles. As mentioned earlier, this chapter selects the canonical ensemble (NVT) and the isothermal–isobaric ensemble (NPT) for molecular simulation calculations. The following sections will provide a detailed description of the Monte Carlo simulation procedures for these two ensembles.

Canonical Ensemble

For the canonical ensemble (NVT ensemble), the Metropolis sampling procedure is as follows:

1. For any particle in the system, apply a displacement with a random direction and a random magnitude.

$$x_{new} = x_{old} + Rand[-1,1] \times \Delta r_{max} \tag{2.30}$$

$$y_{new} = y_{old} + Rand[-1,1] \times \Delta r_{max} \tag{2.31}$$

$$z_{new} = z_{old} + Rand[-1,1] \times \Delta r_{max} \tag{2.32}$$

   In the equation, x_{old}, y_{old}, z_{old} represent the initial positions of the particle in the system space, while x_{new}, y_{new}, z_{new} denote the new positions of the particle after displacement. Δr_{mas} is the maximum displacement set in the simulation.
2. The change in potential energy ΔU of the particle before and after the movement is given by Eq. (2.33).

$$\Delta U = U_{new} - U_{old} \tag{2.33}$$

   In this equation, U_{old} represents the potential energy of the particle before the displacement, and U_{new} represents the potential energy after the displacement.

If $\Delta U \leq 0$, the movement of the particle is accepted with a probability of 1, and the particle's new coordinates become (x_{new},y_{new},z_{new}). If $\Delta U>0$, the movement of the particle is accepted with a probability of $\exp(-\Delta U/k_B T)$, which must be evaluated as shown in Eq. (2.34).

$$\begin{cases} \exp\left(-\dfrac{\Delta U}{k_B T}\right) \geq P, \text{Particle movement is accepted} \\ \exp\left(-\dfrac{\Delta U}{k_B T}\right) < P, \text{Particle movement has no effect} \end{cases} \tag{2.34}$$

In this equation, P is a random number between 0 and 1.

Repeat the above steps to sample all accepted configurations in the system. Based on the accepted configurations, thermodynamic properties and other relevant characteristics are calculated.

Isothermal–Isobaric Ensemble (NPT Ensemble)

For the isothermal–isobaric ensemble (NPT ensemble), the Metropolis sampling procedure is as follows:

1. First, assume that the system in this ensemble is a cubic box containing N particles with side length L. The random displacement of particles within the system or the random change in system volume can be expressed by the following equations:

$$\vec{s}_{i,new} = \vec{s}_{i,old} + \Delta\vec{s}_{\max}\left(2\vec{\xi} - \vec{1}\right) \tag{2.35}$$

$$L_{new} = L_{old} + \Delta L_{\max}\left(2\xi - \vec{1}\right) \tag{2.36}$$

   In the above expressions, ξ is a random number within the range [0,1], and $\vec{\xi}$ is a unit vector indicating the direction of displacement $\left(\vec{\xi}_1, \vec{\xi}_2, \cdots, \vec{\xi}_N\right)$. $\vec{1} = \left(1,1,\cdots,1\right)$; $\vec{s} = \left(s_1, s_2, \cdots, s_N\right)$ represent a set of particle coordinates before and after displacement, respectively. $\vec{s} = L^{-1}\vec{r}$, $L = V^{\frac{1}{3}}$ is the length of the simulation box; and $\Delta\vec{s}_{\max}$ and $\Delta L_{\max}$ are the maximum allowed changes in particle position and box length, respectively.
2. After a particle undergoes a new displacement, the total enthalpy of the system changes. This change is denoted as ΔH, and it can be expressed as follows:

$$\Delta H = \Delta U + P\left(V_{new} - V_{old}\right) - Nk_B T \ln\frac{V_{new}}{V_{old}} \tag{2.37}$$

   In this equation, if $\Delta H = 0$, the displacement is accepted with a probability of 1. If $\Delta H > 0$, the value of $\exp(-\Delta H/k_B T)$ is compared with the random number P, following the same comparison method used in the sampling procedure of the canonical ensemble.

2.3 Construction of a Microscopic Dynamic Displacement Model for Slit Pores in Coal Matrix

In this chapter, the pores within the coal matrix are simplified as slit-shaped pores [158–161]. Based on this simplification, two models are established: a single-component CH_4 gas adsorption/desorption model and a displacement model capable of providing a continuous pressure gradient within slit pores of the coal matrix. These models are used to investigate the transport behavior of CH_4 gas in the coal matrix pores during the gas injection displacement process from the perspective of pore-scale mechanisms.

2.3.1 Construction of the Slit Pore Model in the Coal Matrix

The coal samples used in this study were all collected from the 11–2# coal seam of the Liuzhuang Coal Mine, operated by China Coal Xinji Energy Group. This seam contains bituminous coal. Based on physical and chemical analyses, researchers both domestically and internationally have proposed various molecular structure models for bituminous coal. Among them, the Wiser model is widely regarded as one of the most representative and reasonable models for bituminous coal [101]. The elemental mass percentages in the Wiser model are: 78.3 wt% C, 11.2 wt% O, 5.8 wt% H, 3.2 wt% S, and 1.4 wt% N, which are closely aligned with the composition of real bituminous coal. Therefore, in this chapter, the Wiser bituminous coal molecule is adopted to construct the coal molecular layers on both sides of the slit. The Wiser bituminous coal macromolecular model is shown in Fig. 2.4.

The modeling steps for constructing the coal matrix slit pore model are as follows:

(a) The chemical structure model of the Wiser bituminous coal macromolecule was constructed using the Materials Studio software for preprocessing and model-

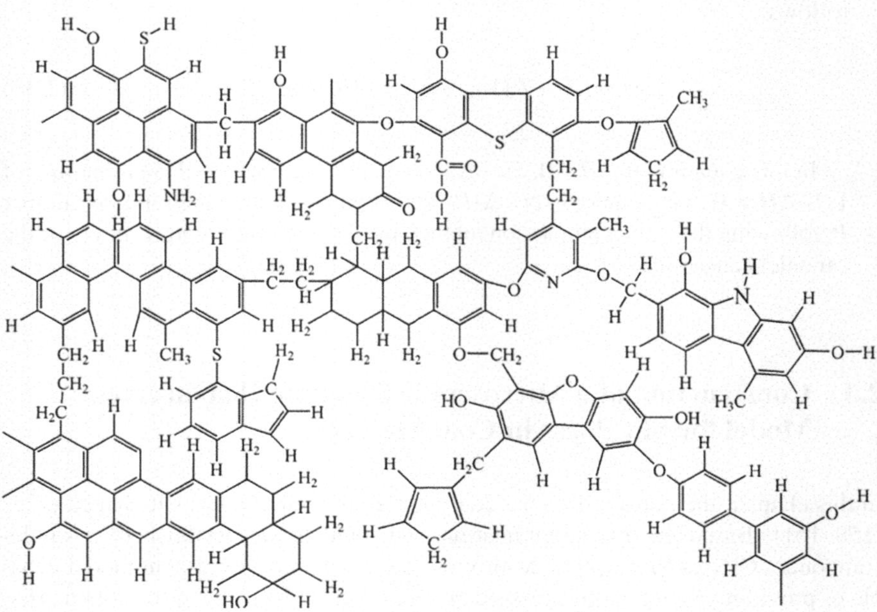

Fig. 2.4 Wiser bituminous coal macromolecular model

ing. All dangling bonds in the macromolecular model were saturated with methyl groups.

(b) The coal molecular layer consists of multiple coal macromolecules. Based on a single macromolecule, a single-layer supercell model of the coal molecular layer was established using the Super Cell module in Materials Studio. Specifically, the number of molecules is 2 in the x-direction, 1 in the y-direction, and 6 in the z-direction, totaling 12 molecules. The Forcite module was used to perform mechanical calculations on the supercell model, with the CVFF force field selected for interaction parameters and atomic charges assigned using the force field assigned method. The non-bonded parameters of the CVFF force field used in this chapter are listed in Table 2.2.

In the table, ε(epsilon) represents the depth of the potential well, i.e., the potential energy value at which the interaction between two atoms or molecules

Table 2.2 Non-bonding parameters of the CVFF force field

Atomic type	Key type	Potential well depth ε(Kcal/mol)	Bond length σ(Å)
C	cp	0.1479999981	3.6170487995
	c*	0.1479999981	3.6170487995
	c3	0.0389999952	3.8754094636
	c2	0.0389999952	3.8754094636
	c1	0.0389999952	3.8754094636
	c=	0.1479999981	3.6170487995
	c CH_4	1.9872e-5	3.31
	c CO_2	0.053649	2.8
	c	0.16	2.5
	c5	0.1479999981	3.6170487995
	cs	0.1479999981	3.6170487995
O	o*	0.2280000124	2.8597848722
	o	0.2280000124	2.8597848722
	oh	0.2280000124	2.8597848722
	op	0.2280000124	2.8597848722
	o CO_2	0.156973	3.05
H	h	0.0380000011	2.4499714540
	ho	0	0
	h CH_4	0.03040	3.31
	hs	0.0380000011	2.4499714540
	h n2	0	0
	hn	0	0
N	np	0.1669999743	3.5012320066
	n=	0.1669999743	3.5012320066
	n2	0.1669999743	3.5012320066
	n n2	0.0715	3.31
S	s	0.0429759086	3.3677544003
	sh	0.0429759086	3.3677544003

is strongest. In the Lennard-Jones potential function, ε indicates the minimum potential energy (or well depth) when two particles are at the optimal distance of attraction. This parameter reflects the strength of the interaction—a larger ε value implies a stronger attractive force. σ(sigma) represents the distance between two atomic nuclei at which the interaction potential energy is zero. In the Lennard-Jones potential function, σ is the distance at which the potential energy transitions from repulsive to attractive. This parameter can be considered the effective diameter of the particle, providing a measure of particle size. A larger σ value indicates a larger effective particle size.

(c) Carbon plates were placed at the top and bottom of the supercell model along the z-direction, followed by vertical compression of the model through NPT relaxation to achieve the target density. The density of the single-layer coal molecular structure was adjusted to 1.25 g/cm^3, which falls within the actual density range of bituminous coal (1.20–1.40 g/cm^3 [162, 163]).

(d) Annealing dynamics simulation was performed on the combined model of carbon plates and the coal molecular layer under the NVT ensemble. The simulation was conducted sequentially at temperatures of 2000, 1000, and 300 K, and the final configuration after reaching thermal equilibrium was exported.

(e) The carbon plate models set in step "c" were deleted, and the single-layer coal molecular structure was rotated by 180° to construct a double-layer coal molecular structure. The completed coal matrix slit pore model is shown in Fig. 2.5. The pore width of the coal matrix was defined by adjusting the distance between the two coal molecular layers.

2.3.2 *Construction of the Single-Component CH_4 Molecular Diffusion Model in Coal Matrix Slit Pores*

Based on the coal matrix slit pore model established in Sect. 2.3.1, a single-component CH_4 gas diffusion model within the coal matrix slit pore is constructed. The modeling steps are as follows:

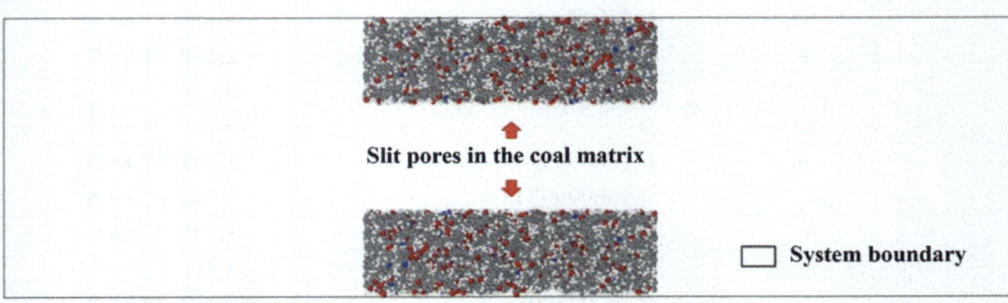

Fig. 2.5 Coal matrix slit pore modeling

2.3.2.1 Construction of the CH_4 Gas Molecule Adsorption Model in Slit Pores

(a) Reduce the system boundaries in the x-direction to the edges of the coal macromolecular layers on both sides of the slit, in order to ensure that the gas molecules only adsorb within the slit and on the surface of the coal matrix;
(b) Introduce a certain number of monomeric CH_4 gas molecules into the slit, read the combined model constructed in this step along with the potential parameters for all atom types, iteratively adjust the atomic coordinates, perform energy minimization on the system, and export the optimized configuration. For CH_4 molecules, the TraPPE-EH force field [164] is adopted, treating the gas molecules as rigid bodies with fixed bond lengths and bond angles. The potential parameters are shown in Table 2.3.
(c) Read the energy-minimized configuration exported from step "b" and perform NVT relaxation on the gas molecules within the slit. The relaxation process corresponds to the CH_4 gas adsorption process (Fig. 2.6).

2.3.2.2 Construction of the CH_4 Gas Molecule Desorption Model within the Slit Pore

(a) Based on the previously established single-component CH_4 gas adsorption model within the coal matrix slit pore, expand the system area on the right side of the slit pore, as shown in Fig. 2.7.
(b) In the newly generated system, the CH_4 gas molecules within the slit pore are subjected to NVT relaxation until the CH_4 molecules complete the NVT relaxation process within the newly expanded system space, thereby reaching desorption equilibrium.

Table 2.3 Lennard-Jones 12–6 potential energy parameters and atomic charges for the CH_4 molecule

Atomic type	ε(Kcal/mol)	σ(Å)	q(e)
[C]H4	0.01	3.31	0.000
H-[M]-C	15.30	3.31	0.000
Bond length	b_0(Å)	Bond angle	θ_0(°)
C-M	0.55	H-C-H	109.5

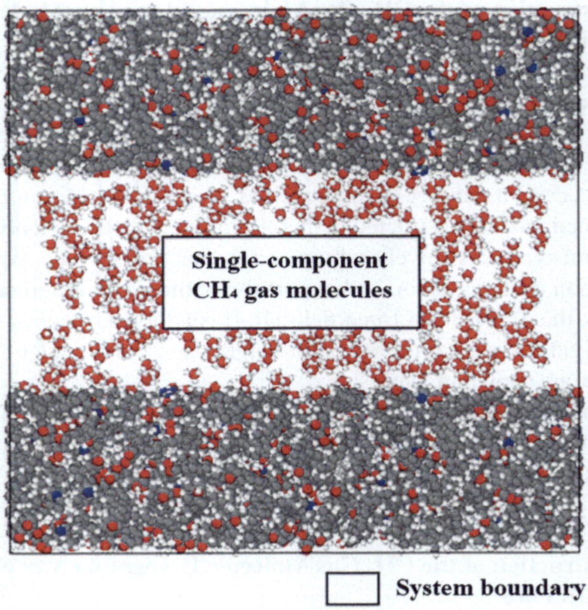

Fig. 2.6 Microscopic CH_4 gas adsorption model for coal matrix slit pores

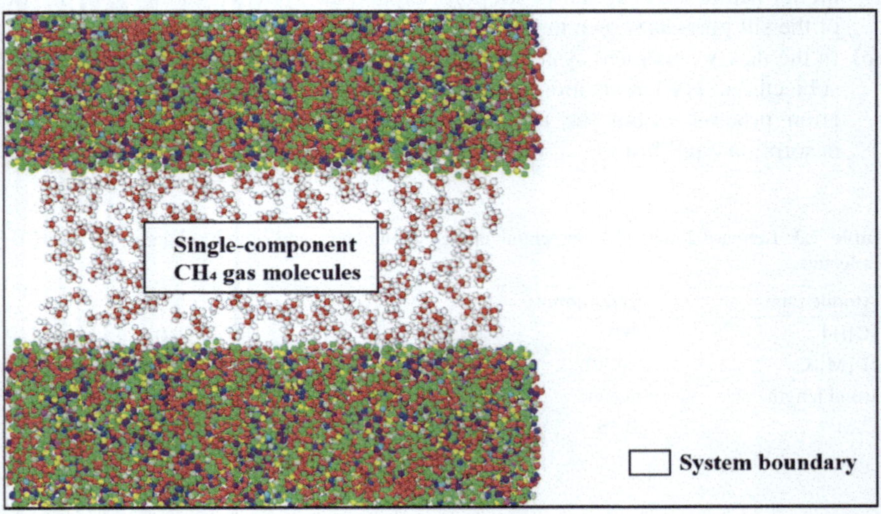

Fig. 2.7 Microscopic CH_4 gas desorption model for coal matrix slit pores

2.3.3 Construction of the Dynamic Displacement Model for Slit Pores in Coal Matrix

At the macroscopic scale, gas injection displacement is considered an effective method for enhancing the production of coalbed methane (CBM) and shale gas. Common injection gases include N_2 and CO_2, which act through different mechanisms. CO_2 has a stronger adsorption capacity than CH_4 and can displace adsorbed CH_4 in coal pores via competitive adsorption. In contrast, N_2 has a weaker adsorption capacity than CH_4; therefore, the conventional view holds that N_2 injection promotes CH_4 desorption mainly by reducing the partial pressure of methane in the coal seam. At the pore scale, the gas injection displacement process in coal is characterized by the interactions between injected gas molecules (N_2 or CO_2) and CH_4 molecules within the coal matrix pore channels, and their directional flow under a pressure gradient. To simulate this process, a gas injection displacement model was developed based on the slit pore coal matrix model constructed in Sect. 2.3.1. This model consists of three components: the injection zone for heterogeneous gas, the slit pore region, and the back-pressure zone. The model construction steps are as follows:

(a) Carbon plates are placed on both sides of the slit pore, with a designated number of CO_2 or N_2 molecules positioned in the space between the left carbon plate and the left side of the slit. Meanwhile, a certain number of CH_4 molecules are placed inside the slit pore to form a composite model. The purpose of introducing carbon plates is to generate a driving force, enabling CO_2 or N_2 molecules to enter the slit under a defined pressure gradient, thereby simulating a dynamic displacement process with a continuous pressure gradient. For CO_2 and N_2 molecules, the TraPPE-EH force field is adopted. The gas molecules are treated as rigid bodies with fixed bond lengths and bond angles. The Lennard-Jones 12–6 potential parameters and atomic charges for each molecule are listed in Tables 2.4 and 2.5, respectively.
(b) Read the composite model and all atomic potential parameters constructed in step "a." Perform iterative adjustments to the atomic coordinates to minimize the system energy, and export the energy-minimized configuration.
(c) Load the energy-minimized configuration from step "b," then sequentially perform NVT relaxation for the CH_4 and CO_2/N_2 molecules placed in step "a," and export the resulting NVT-relaxed composite model.

Table 2.4 Lennard-Jones 12–6 potential energy parameters and atomic charges of CO_2 molecules

Atomic type	ε (Kcal/mol)	σ (Å)	q(e)
O=[C]=O	27.0	2.800	0.700
[O]=C=O	79.0	3.050	−0.350
Bond length	b_0(Å)	Bond angle	θ_0(°)
O=(C=O)	1.16	O = (C) = O	180.00

Table 2.5 Lennard-Jones 12–6 potential energy parameters and atomic charges of N_2 molecules

Atomic type	ε (Kcal/mol)	σ (Å)	q(e)
[N]#N	36.0	3.310	−0.482
N#[M]#N	0.00	0.000	0.964
Bond length	b_0(Å)	Bond angle	θ_0(°)
N#N	1.10	N#(M)#N	180.00
N#M	0.55		

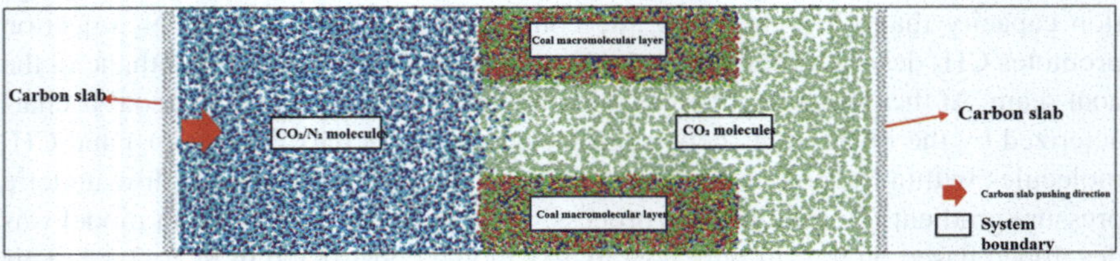

Fig. 2.8 Coal matrix slit pore micro-excavation modeling

(d) Load the composite model from step "c," apply a specific force to push the left carbon plate to generate a continuous pressure gradient, thereby simulating the process of CO_2/N_2 injection for CH_4 displacement in coal. During the displacement process, the number and density of molecules in the slit are recorded to inversely calculate the injection amount, adsorption amount, and desorption amount of gas molecules. The coal matrix slit pore displacement model is shown in Fig. 2.8.

2.4 Adsorption and Desorption Characteristics of Pure CH_4 Molecules in Coal Matrix Slit Pores

Based on the pure CH_4 gas adsorption model in coal matrix slit pores established in Sect. 2.3.2, the adsorption characteristics of pure CH_4 gas in coal matrix slit pores under different temperatures and slit widths were investigated.

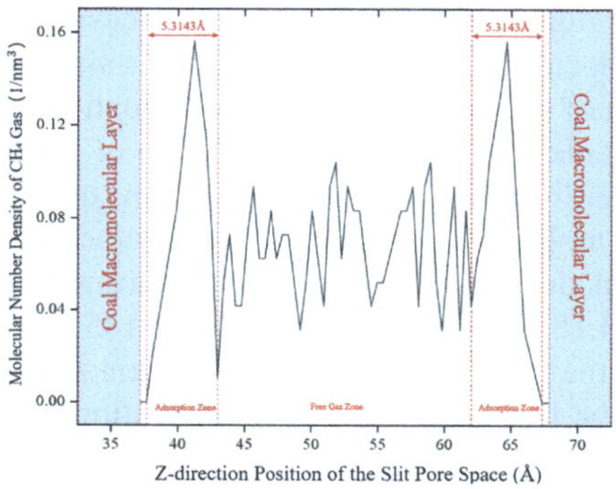

(a) 298.15 K

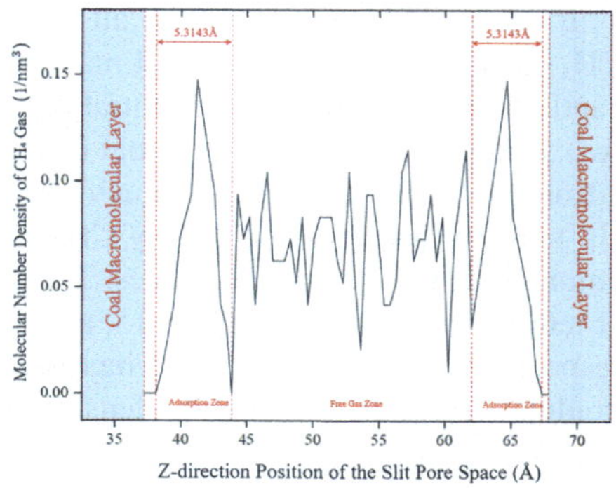

(b) 323.15 K

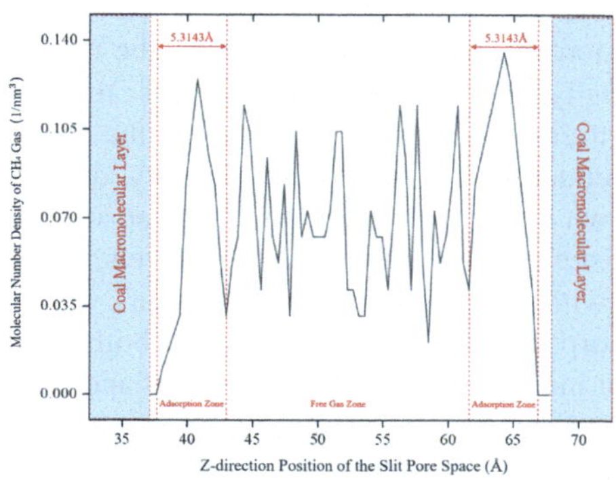

(c) 353.15 K

Fig. 2.9 Density distribution in *z*-direction of CH_4 molecules stabilized by adsorption at various temperatures. (**a**) 298.15 K. (**b**) 323.15 K. (**c**) 353.15 K

2.4.1 Adsorption Characteristics of Single CH_4 Molecules in Coal Matrix Slit Pores

2.4.1.1 Effect of Temperature on the Adsorption Characteristics of Single CH_4 Molecules

To investigate the adsorption characteristics of single CH_4 gas molecules in slit pores at different temperatures, three temperature levels were selected: 298.15, 323.15, and 353.15 K. The slit pore width was set to 3 nm (with the pore dimensions being 80 Å × 40 Å × 30 Å in length × width × height), and the total number of CH_4 molecules within the slit pore was 425. Figure 2.9 shows the gas density distribution along the *z*-direction within the slit pore at equilibrium adsorption under different temperature conditions.

It can be observed that under all temperature conditions, the number density of CH_4 gas molecules near the coal macromolecular layers in the slit pores is significantly higher than that in the central region of the pore, indicating the formation of two adsorption zones near the coal macromolecular layers. The formation of these adsorption zones is due to the much stronger interactions between the atoms on the slit pore surfaces and the nearby CH_4 gas molecules compared to the interactions among CH_4 molecules themselves. The peak density values in the adsorption zones at temperatures of 298.15, 323.15, and 353.15 K are 0.156, 0.147, and 0.135 nm^3, respectively. As the temperature increases, the peak density in the adsorption zones gradually decreases, while the density in the free region increases, indicating that the adsorption strength of the coal macromolecular layers for CH_4 molecules weakens with rising temperature. This suggests that higher temperatures are unfavorable for the adsorption of CH_4 molecules in slit pores. This phenomenon can be attributed to the fact that the adsorption of CH_4 by coal is an exothermic process. As the temperature rises, more heat is released, causing the equilibrium to shift toward desorption. Additionally, higher temperatures increase the thermal motion of the gas molecules, making it easier for them to overcome the adsorption forces from the coal matrix, thus reducing the adsorption capacity of the pore walls for CH_4. Furthermore, as shown in Fig. 2.9, the width of the adsorption zones within the slit pores remains consistent across all temperature levels, with a value of 5.3143 Å, which is consistent with the findings reported in Refs. [111, 165]. This is because the width of the adsorption zones is determined by both the distribution of adsorption sites on the coal macromolecular layer surfaces and the strength of their interaction with CH_4 molecules. The adsorption process involves CH_4 molecules occupying these fixed adsorption sites, the number of which is constant and unaffected by temperature. Therefore, the width of the adsorption zones does not change. However, the number of CH_4 molecules adsorbed within these zones can be influenced by temperature and other conditions. As the temperature increases, fewer CH_4 molecules are retained at the adsorption sites, leading to a decrease in adsorption density. Reference [165] also points out that under different pressure conditions, the adsorption layer width for the same gas molecules remains unchanged.

Based on the aforementioned characteristics, this chapter quantifies the amount of gas molecules adsorbed and in the free state under different conditions by measuring the adsorption zone and free region widths within the slit pores. This provides a basis for further investigation into the dynamic migration behavior of gas molecules in coal slit pores. Figure 2.10 shows the changes in final adsorption and free-state quantities of CH_4 molecules within coal matrix slit pores under different temperature conditions. It is evident that the CH_4 adsorption amount in the slit pores is negatively correlated with temperature, while the CH_4 free-state amount is positively correlated with temperature.

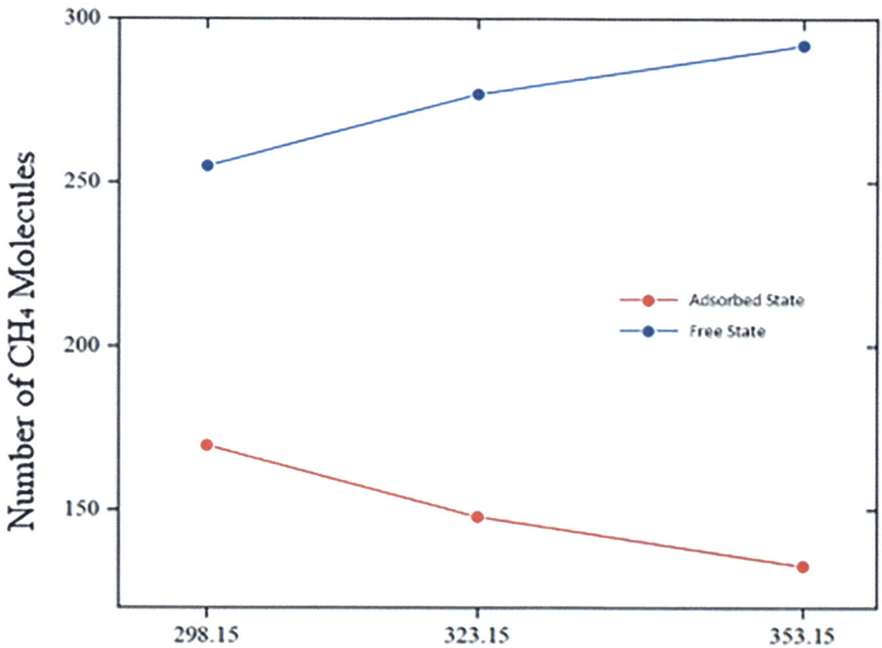

Fig. 2.10 Number of CH_4 molecules in adsorbed and free state at each temperature level condition

2.4.1.2 Effect of Slit Width on the Adsorption Characteristics of Single CH_4 Molecules

Coal is a porous medium with a complex matrix of pores and fractures, where gas exists in two forms: free and adsorbed. Adsorbed gas accounts for as much as 80–90% of the total gas content in coal seams [166, 167]. The pores in the coal matrix, characterized by their large specific surface area, provide numerous adsorption sites for the adsorbed gas. This section adopts the classification of coal matrix pore sizes proposed by the International Union of Pure and Applied Chemistry (IUPAC) [168]. According to the classification, coal pores are categorized by diameter into micropores (<2 nm), mesopores (2–50 nm), and macropores (>50 nm). Micropores serve primarily as adsorption sites for gas molecules, mesopores act as diffusion spaces, and macropores function as the main channels for gas flow. Together, micropores and mesopores form the principal storage space for gas in coal. In this study, slit-shaped pores with widths of 1, 1.5, 3, and 6 nm were selected to represent micropores (1, 1.5 nm) and mesopores (3, 6 nm). The adsorption behavior of CH_4 gas molecules within these slit pores was simulated at the pore scale to investigate the effect of slit width on CH_4 adsorption characteristics. The molecular configurations used in the simulations are shown in Fig. 2.11. The simulation temperature was set at 298.15 K, and 425 CH_4 molecules were placed in the slit pores.

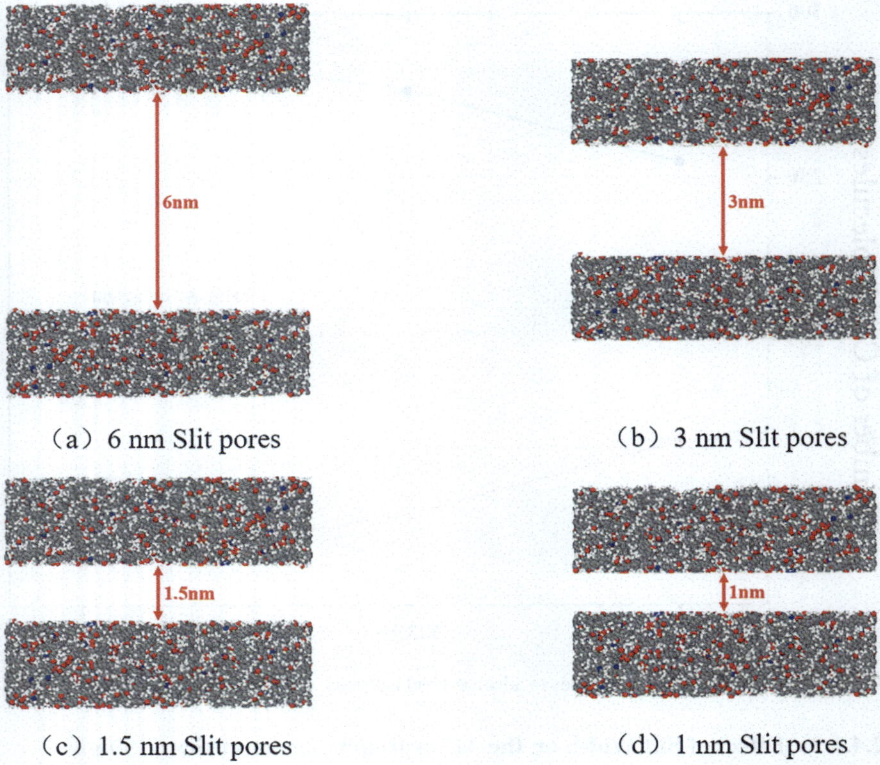

Fig. 2.11 Slit pore configuration. (**a**) 6 nm slit pores. (**b**) 3 nm slit pores. (**c**) 1.5 nm slit pores. (**d**) 1 nm slit pores

Snapshots of adsorption equilibrium for the four pore widths are presented in Fig. 2.12.

To better observe the distribution of CH_4 gas molecules within slit pores of different widths, the relative molecular number density was used for comparison. This density represents the proportion of CH_4 molecules located on the *xy*-plane at a specific *z*-axis position relative to the total number of CH_4 molecules within the slit pore. The relative number density distribution curves of CH_4 gas molecules at adsorption equilibrium are shown in Fig. 2.13. The adsorption zone width and the numbers of adsorbed and free-state CH_4 molecules under each slit pore width condition are presented in Fig. 2.14. It can be observed that CH_4 molecules accumulate on the surface of the coal macromolecular layers, forming strong adsorption zones on both sides of the slit. Since the number of CH_4 gas molecules placed in the slit pores is the same across different widths, narrower slit widths result in higher internal gas pressures. This leads to an increase in the adsorption of CH_4 molecules by the coal macromolecular layers, with a corresponding increase in the number of adsorbed CH_4 molecules and a decrease in the number of free CH_4 molecules.

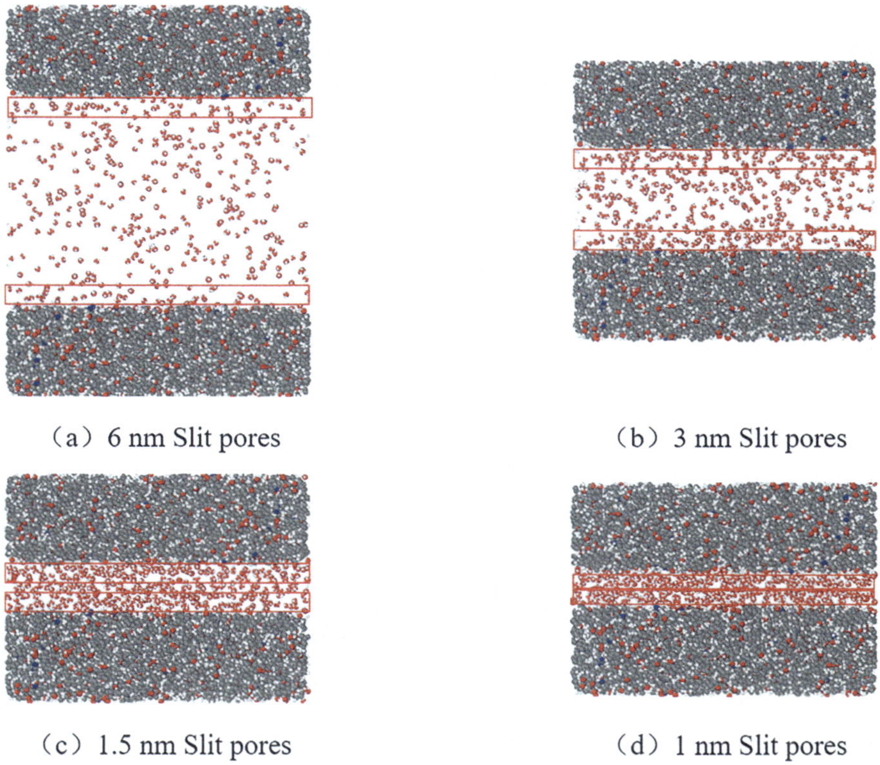

Fig. 2.12 Snapshot of molecular adsorption equilibrium of CH_4 gas. (**a**) 6 nm slit pores. (**b**) 3 nm slit pores. (**c**) 1.5 nm slit pores. (**d**) 1 nm slit pores

Notably, when the slit width is 1 nm, the two adsorption peaks on either side are continuous, indicating the absence of a free gas region. As the pore narrows, the confined space tends to "squeeze" the molecules into a very limited volume, leaving no room for free-state CH_4 molecules. Similar phenomena have been reported in the literature [108, 111]. Moreover, it is also observed that as the slit width decreases, the width of the CH_4 adsorption zones gradually increases. The adsorption zone widths for 6, 3, and 1.5 nm slit pores are 4.5601, 5.3143, and 6.4716 Å, respectively, while the entire 1 nm pore is an adsorption zone. This phenomenon arises mainly because the adsorption process in coal slit pores involves interactions between gas molecules and the pore walls [169]. When the pore width is small, molecules are more likely to interact with the pore walls due to their proximity, causing them to stay near the wall for longer periods and thus broadening the adsorption zone. As the pore width increases, this interaction weakens, leading to a narrower adsorption region. Therefore, narrower slit pores in the coal matrix result in wider adsorption zones along the pore walls.

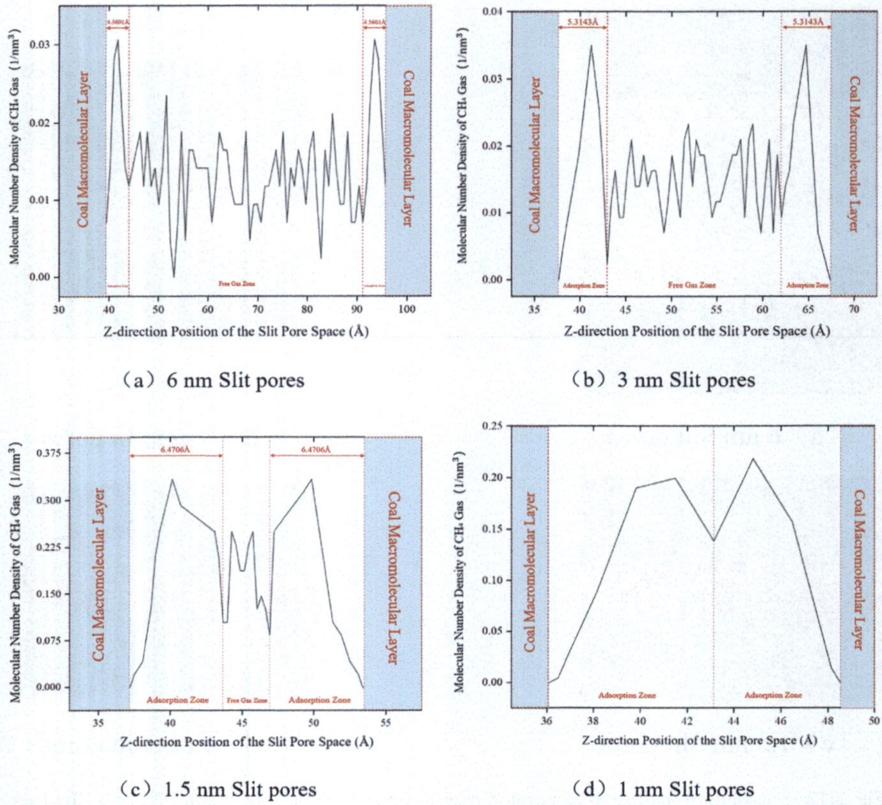

Fig. 2.13 Gas density distribution in the *z*-direction for CH_4 molecule adsorption stabilization under each slit pore width condition. (**a**) 6 nm slit pores. (**b**) 3 nm slit pores. (**c**) 1.5 nm slit pores. (**d**) 1 nm slit pores

2.4.2 Analysis of Desorption Characteristics of Single CH_4 Molecules in Coal Matrix Slit Pores

In this section, molecular simulations of CH_4 gas desorption and diffusion under adsorption equilibrium within slit pores are conducted to model the "pressure-reduction extraction" process used in coalbed methane (CBM) engineering and the "conventional discharge" method applied in underground gas disaster mitigation projects. The study considers the effects of temperature and slit pore width on the desorption behavior of CH_4 molecules, providing essential data support for subsequent comparative analyses with the dynamic diffusion simulations of CH_4 during gas injection displacement. The temperature levels, slit pore widths, and molecular quantities within the slits are consistent with those used in Sect. 2.4.1. When examining the influence of temperature on desorption characteristics, the

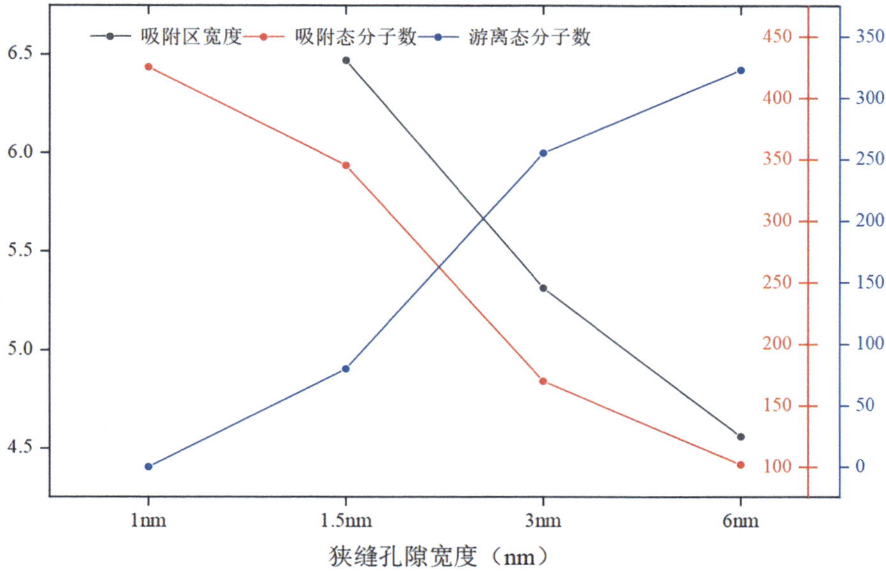

Fig. 2.14 Width of adsorption zone, number of CH_4 molecules in adsorbed and free states for each slit pore width condition

simulation is set to run for a total of 220,000 steps; when investigating the effect of slit pore width, the simulation runs for 500,000 steps, with a time step of 1 fs per step.

2.4.2.1 Effect of Temperature on the Desorption Characteristics of Single CH_4 Molecules

Figure 2.15 shows the variation in the number of adsorbed CH_4 molecules, the cumulative desorbed amount, and the emitted amount of CH_4 molecules within the slit pores during the desorption process under different temperature conditions.

It can be observed that during the desorption process, the number of adsorbed CH_4 molecules in the slit pores initially decreases rapidly, then gradually slows down, and finally stabilizes. Meanwhile, the cumulative desorption amount of CH_4 molecules in the slit pores shows a trend of initially rapid increase, followed by a slower increase, and eventually stabilizes. At the end of the pressure-reduction desorption process, the cumulative desorption amounts of adsorbed CH_4 at temperature levels of 298.15, 323.15, and 353.15 K are 57, 62, and 68 molecules, respectively. This indicates that as the system temperature increases, the desorption rate of adsorbed CH_4 molecules in the slit pores accelerates, and the total cumulative desorption amount also increases accordingly. With the rise in temperature, the

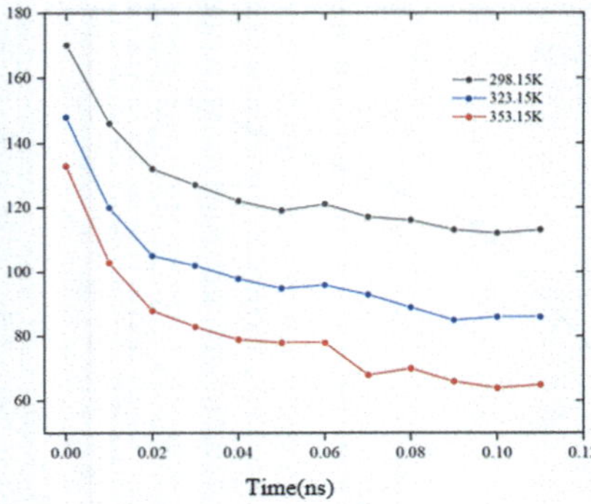

(a) Molecular weight of CH4 in the adsorbed state

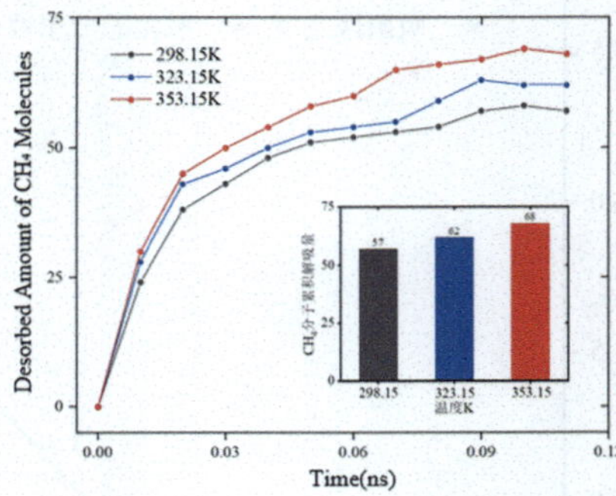

(b) Cumulative desorption of CH4 molecules

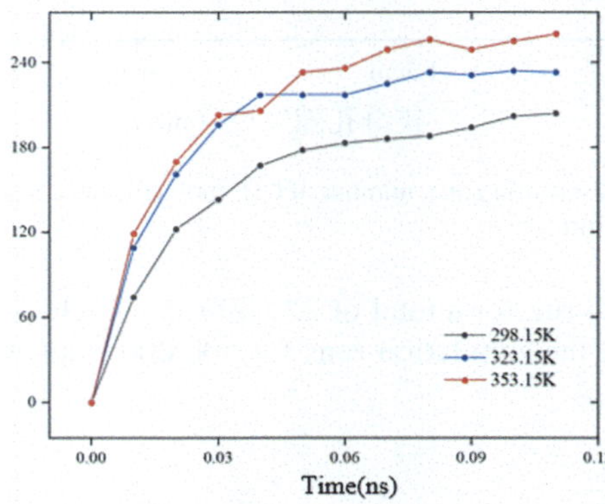

(c) CH4 molecular emissions

Fig. 2.15 Changes in molecular weight of CH_4 in adsorbed state, cumulative desorption of CH_4 molecules, and emissions of CH_4 molecules within slit pores. (**a**) Molecular weight of CH_4 in the adsorbed state. (**b**) Cumulative desorption of CH_4 molecules. (**c**) CH_4 molecular emissions

kinetic energy of CH_4 molecules within the slit pores increases, and the thermal motion of CH_4 molecules becomes more intense [170], making it easier for the CH_4 molecules initially adsorbed on the slit pore surfaces to desorb. Furthermore, the higher the system temperature, the faster the desorption rate and the greater the cumulative desorption amount.

Additionally, the desorption of CH_4 molecules within the slit pores is governed by the concentration gradient, while the transport of CH_4 molecules between the inside and outside of the slit pores is controlled by the pressure gradient. As desorption proceeds, the concentration gradient of CH_4 molecules between the pore wall surface and the free region inside the slit gradually decreases, leading to a progressive reduction in the CH_4 desorption rate until equilibrium is reached. Similarly, the pressure gradient between the inside and outside of the slit pores also diminishes over time, causing the number of CH_4 molecules in the exhaust region to increase rapidly in the early stage, then slow down, and eventually stabilize. At the end of the desorption process, the number of CH_4 molecules in the back-pressure region at temperature levels of 298.15, 323.15, and 353.15 K is 204, 233, and 260 molecules, respectively, accounting for 48.0%, 54.8%, and 61.2% of the total number of CH_4 molecules in the system. From the pore-scale perspective, elevated temperature promotes the desorption of coalbed methane. However, the duration from the onset of CH_4 desorption to the point of desorption equilibrium shows no significant difference across the different temperature conditions.

2.4.2.2 Effect of Slit Pore Width on the Desorption Characteristics of Single CH_4 Molecules

Figure 2.16 illustrates the evolution of the number of adsorbed CH_4 molecules, cumulative desorption, and CH_4 release during the desorption process under different slit pore width conditions. As shown in Fig. 2.16a, b, regardless of slit width, the amount of CH_4 desorbed increases rapidly at first, then more gradually, and eventually reaches a plateau over time. At the same temperature, decreasing slit pore width leads to both a higher initial desorption rate and a greater final cumulative desorption. As discussed in Sect. 2.4.1, for the same total number of CH_4 molecules, narrower slit pores result in more CH_4 being adsorbed. Consequently, a steeper concentration gradient forms between the adsorbed CH_4 near the pore walls and the free CH_4 within the pore space, thereby driving a faster early-stage desorption rate and ultimately a higher cumulative desorption under pressure-reduction conditions. At the end of desorption, the cumulative numbers of desorbed CH_4 molecules for slit pore widths of 1, 1.5, 3, and 6 nm are 251, 187, 69, and 48, respectively. However, even after desorption has ceased, a considerable number of CH_4 molecules remain in the adsorbed state—specifically 174, 158, 101, and 60 molecules for the same respective pore widths—indicating that narrower pores retain more adsorbed CH_4. Figure 2.16c further shows that the number of CH_4 molecules released into the back-pressure region increases with slit pore width. The total number of CH_4 molecules released for slit widths of 1, 1.5, 3, and 6 nm is 248, 253, 259, and 273, respectively, corresponding to 58.4%, 59.5%, 60.9%, and 64.2% of the total CH_4 molecules in the system. Larger slit widths result in both a higher overall release and a faster early-stage release rate. Moreover, with increasing slit width, a growing proportion of the released CH_4 originates from the free (non-adsorbed) state within the pore. The contribution of free-state CH_4 to total desorbed CH_4 increases with pore width, accounting for 100%, 73.9%, 26.6%, and 17.6% of the total released CH_4 for 1, 1.5, 3, and 6 nm pores, respectively. These results indicate that narrower pores enhance CH_4 adsorption and desorption rates but retain more residual adsorbed CH_4, while wider pores promote more efficient gas release, primarily from the free state.

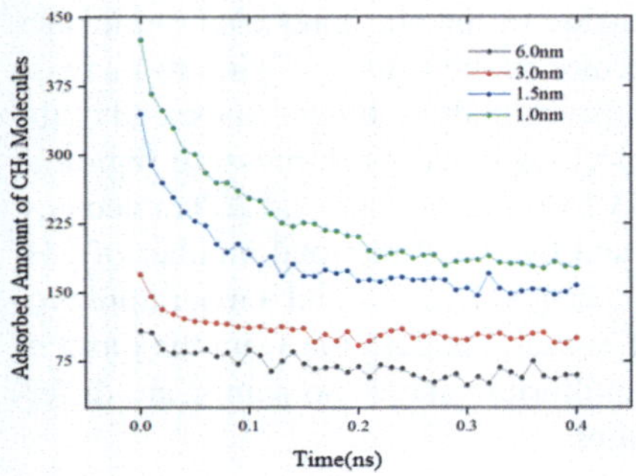

(a) Molecular weight of CH_4 in the adsorbed state

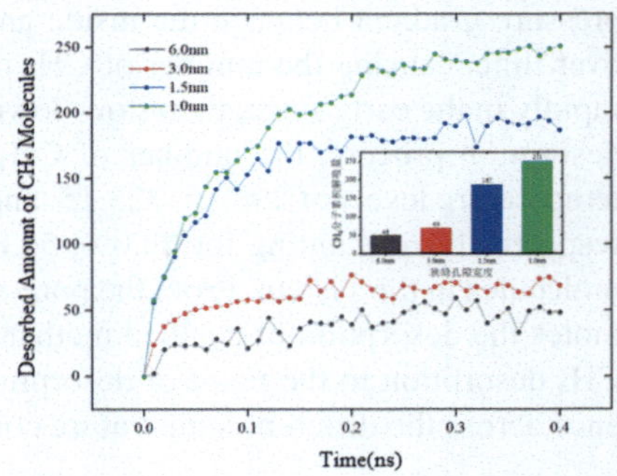

(b) Cumulative desorption of CH_4 molecules

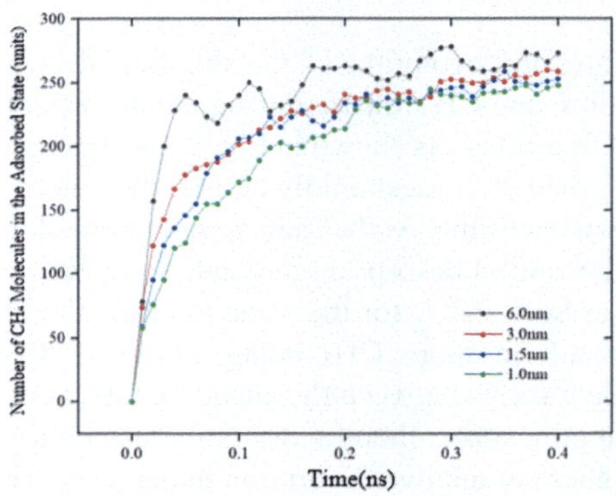

(c) CH_4 molecular emissions

Fig. 2.16 Changes in molecular weight of CH_4 in adsorbed state, cumulative desorption of CH_4 molecules, and emissions of CH_4 molecules within slit pores. (**a**) Molecular weight of CH_4 in the adsorbed state. (**b**) Cumulative desorption of CH_4 molecules. (**c**) CH_4 molecular emissions

Based on the above analysis, it can be concluded that, at the pore scale, CH_4 recovery under conventional "pressure-reduction extraction" or "standard degassing" techniques primarily originates from mesopores and larger pores. These pore types contribute mainly through the recovery of free-state CH_4 molecules. In contrast, micropores predominantly store CH_4 in the adsorbed state, contributing less directly to CH_4 production under such conditions.

2.5 Microscopic Characteristics of CH_4 Molecular Transport during Gas Injection Displacement in Coal Matrix Slit Pores

Based on the dynamic displacement model of coal matrix slit pores established in Sect. 2.3, the total number of CH_4 molecules initially present in the slit pores is defined as $N_{pore\text{-}CH4}$. During the displacement process, the number of CH_4 molecules in the adsorption region and the free region of the slit pores are defined as $N_{ad\text{-}CH4}$ and $N_{free\text{-}CH4}$, respectively. The amount of desorbed CH_4 is denoted as $N_{de\text{-}CH4}$. The total number of heteronuclear gas molecules (CO_2 or N_2) in the slit pores during displacement is defined as $N_{pore\text{-}CO2/N2}$, which is the sum of the number of heteronuclear molecules in the adsorption region ($N_{ad\text{-}CO2/N2}$) and in the free region ($N_{free\text{-}CO2/N2}$). In addition, the increase in CH_4 molecules in the back-pressure region is defined as the CH_4 displacement amount (N_{dis}), and the increase in heteronuclear gas molecules in the back-pressure region is defined as $N_{back\text{-}CO2/N2}$.

The recovery rate R_{dis} of CH_4 from slit pores under gas injection and production conditions is calculated using Eq. (2.38):

$$R_{dis} = N_{dis} / N_{pore-CH_4} \times 100\% \tag{2.38}$$

2.5.1 *Effect of Temperature on the Dynamic Migration Behavior of CH_4 Molecules*

To investigate the influence of different temperature levels on the dynamic migration behavior of CH_4 gas molecules during the gas injection displacement process in coal, simulations were conducted at three temperatures: 298.15, 323.15, and 353.15 K. The slit pore width was set to 3 nm, and the injected heterogeneous gases were CO_2 and N_2, respectively. A total of 1000 CH_4 molecules were initially equilibrated under NVT conditions across both the pore region and the back-pressure region to ensure system stability. During the displacement simulation, a rightward driving force of 2×10^{-5} kcal/(mol·Å) was applied to the left carbon plate to generate a continuous pressure gradient. The total simulation duration was set to 400,000 steps with a time step of 1 fs per step.

2.5.1.1 Variation in CH_4 Adsorption Within Slit Pores

Figure 2.17 illustrates the temporal evolution of the number of adsorbed CH_4 molecules ($N_{ad\text{-}CH4}$) within the adsorption region of the slit pore under different temperature conditions during the CO_2/N_2 injection displacement process. Overall, a distinct difference is observed between the two injection gases. When N_2 is used as the

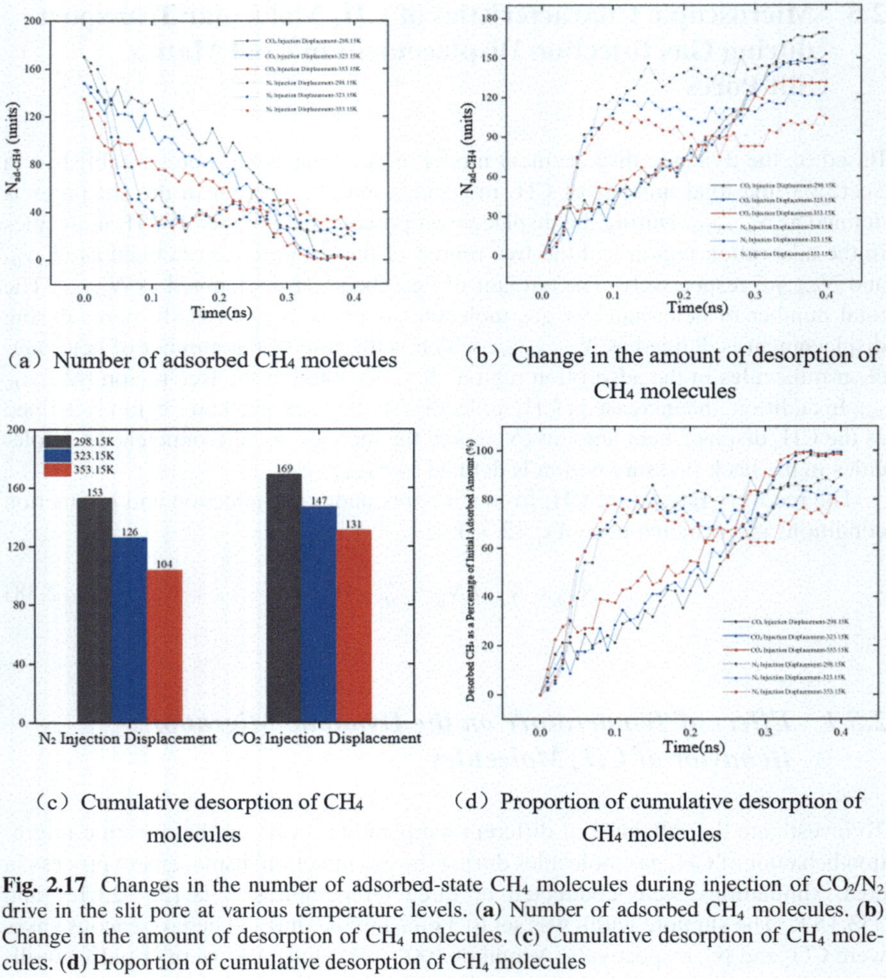

(a) Number of adsorbed CH_4 molecules (b) Change in the amount of desorption of CH_4 molecules

(c) Cumulative desorption of CH_4 molecules (d) Proportion of cumulative desorption of CH_4 molecules

Fig. 2.17 Changes in the number of adsorbed-state CH_4 molecules during injection of CO_2/N_2 drive in the slit pore at various temperature levels. (**a**) Number of adsorbed CH_4 molecules. (**b**) Change in the amount of desorption of CH_4 molecules. (**c**) Cumulative desorption of CH_4 molecules. (**d**) Proportion of cumulative desorption of CH_4 molecules

injection source, the adsorbed CH_4 molecules in the slit pore exhibit a two-stage desorption process: an initial rapid decline followed by a slower, gradual desorption phase. In contrast, when CO_2 is used as the injection gas, the desorption of adsorbed CH_4 molecules proceeds at a relatively steady rate throughout the entire displacement process until desorption reaches completion. This behavior highlights the different displacement mechanisms of the two gases: N_2 mainly reduces partial pressure to promote CH_4 desorption, resulting in a faster initial release, while CO_2, due to its stronger adsorption capacity, displaces CH_4 through a continuous competitive adsorption mechanism.

Specifically, when N_2 is injected, within the first 0.1 ns, the proportion of desorbed CH_4 relative to the initial adsorbed amount follows the order: 353.15 K > 323.15 K > 298.15 K. This trend arises because higher temperatures increase the thermal motion of molecules, making initially adsorbed CH_4 more prone to desorption under the influence of N_2 partial pressure. By the end of this stage, the number of desorbed CH_4 molecules at 298.15, 323.15, and 353.15 K is 102, 108, and 120, respectively, accounting for 78.43%, 85.71%, and 98.08% of the initial adsorbed amount. These results indicate that higher temperatures promote more efficient CH_4 desorption and lead the system closer to dynamic adsorption–desorption equilibrium. However, after 0.1 ns, the trend reverses, with the order becoming: 298.15 K > 323.15 K > 353.15 K. This indicates that at lower temperatures, CH_4 desorption continues at a more sustained rate. With longer displacement durations, lower temperature conditions enable greater CH_4 desorption. Moreover, at higher temperatures, re-adsorption of CH_4 is observed within the slit pores, though the amount of re-adsorbed CH_4 is minimal. These observations suggest that in N_2 injection displacement processes, selecting an appropriate displacement duration based on temperature is essential for effectively reducing CH_4 adsorption in coal seams while optimizing both economic and time efficiency. In contrast, for CO_2 injection, the CH_4 desorption rate remains relatively constant over time across all temperature levels. At higher temperatures, a slight decrease in desorption rate is observed. This indicates that, under CO_2 injection conditions in micropores, the relationship between temperature and CH_4 desorption rate is inversely proportional.

As shown in Fig. 2.17, under N_2 injection, the total number of desorbed CH_4 molecules in the slit pores at temperatures of 298.15, 323.15, and 353.15 K is 153, 126, and 104, respectively, corresponding to 90.00%, 85.14%, and 78.20% of the initially adsorbed CH_4. Under CO_2 injection, the desorption totals are 169, 147, and 131 CH_4 molecules, accounting for 99.41%, 99.32%, and 98.50% of the initially adsorbed CH_4, respectively. These results indicate that, regardless of whether N_2 or CO_2 is used as the injection gas, increasing the temperature is detrimental to the desorption of adsorbed CH_4 molecules within the slit pores.

2.5.1.2 Displacement and Recovery Ratio of CH_4 Molecules in Slit Pores

Figure 2.18 illustrates the variation in CH_4 displacement quantity and recovery ratio within the slit pores over time during CO_2/N_2 injection-driven displacement at different temperature levels.

It can be observed that, regardless of whether N_2 or CO_2 is used as the injection gas, the displacement quantity of CH_4 molecules increases progressively with rising temperature. For N_2 injection, as the temperature increases from 298.15 to 323.15 K, the CH_4 displacement quantity increases by 17 molecules, with the recovery ratio increasing by 4.00%. When the temperature further rises from 323.15 to 353.15 K, the displacement quantity increases by 11 molecules, and the recovery ratio increases by 2.59%. Similarly, under CO_2 injection, as the temperature increases from 298.15 to 323.15 K, the displacement quantity increases by 14 molecules and the recovery ratio increases by 3.29%; when the temperature increases from 323.15 to 353.15 K, the displacement quantity increases by 7 molecules, with the recovery ratio rising by 1.65%. These results indicate that increasing the injection

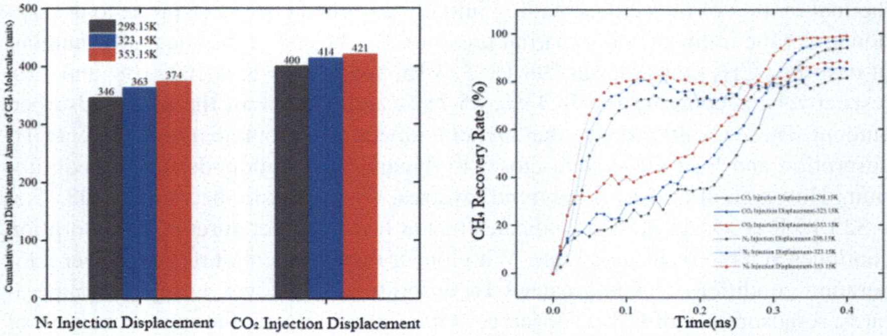

(a) CH_4 displacement molecular weight (b) CH_4 molecular recovery

Fig. 2.18 Variation of CH_4 drive volume and recovery rate in the slit during injection of CO_2/N_2 drive at each temperature level. (**a**) CH_4 displacement molecular weight. (**b**) CH_4 molecular recovery

temperature can enhance the CH_4 displacement quantity and recovery ratio within the slit pores, though the improvement is relatively modest. The detailed CH_4 displacement quantities and recovery ratios under CO_2/N_2 injection at different temperature levels are summarized in Table 2.6.

As shown in Fig. 2.18, higher temperatures not only lead to a faster initial rate of CH_4 displacement but also shorten the time required to reach displacement equilibrium. This phenomenon can be attributed to two main factors. First, as discussed in Sect. 2.4.2, an increase in temperature enhances the thermal motion energy of CH_4 molecules, thereby promoting their spontaneous desorption from the slit pore walls. Second, elevated temperatures improve the interaction between the injected heteronuclear gas molecules and the pore surfaces, which increases the efficiency of CH_4 displacement. As a result, CH_4 molecules are more readily driven out of the slit pores under higher temperature conditions.

2.5.2 *Effect of Slit Pore Width on the Dynamic Migration Behavior of CH_4 Molecules*

To investigate the influence of slit pore width on the dynamic migration behavior of CH_4 molecules during gas injection displacement in coal, two representative slit pore widths—3 and 1.5 nm—were selected to represent mesopores and micropores, respectively. The temperature was maintained at 298.15 K, and the injected gas species included CO_2 and N_2. To ensure that the initial number of stably adsorbed CH_4 molecules within the 3 and 1.5 nm slit pores was consistent, the number of CH_4 molecules was chosen to match the setup described in Sect. 2.4.2, where the effect of pore width on CH_4 desorption characteristics was analyzed. A driving force of 2×10^{-5} kcal/(mol·Å) was applied to the left carbon wall to generate a continuous pressure gradient, and the total simulation time was set to 400,000 steps with a time step of 1 fs.

Table 2.6 CH_4 molecular replacement and recovery in slit pores

	298.15 K		323.15 K		353.15 K	
Temperature	Displacement	Recovery rate	Displacement	Recovery rate	Displacement	Recovery rate
CO_2 displacement	400	94.12%	414	97.41%	421	99.06%
N_2 displacement	346	81.41%	363	85.41%	374	88.00%

2.5.2.1 Variation in CH_4 Adsorption Amount Within Slit Pores

Figure 2.19 presents the variation in the number of adsorbed CH_4 molecules over time during the CO_2/N_2 injection displacement process for slit pores with widths of 1.5 and 3 nm.

It can be observed that regardless of whether N_2 or CO_2 is injected, the desorption rate of CH_4 molecules in the 1.5 nm slit pores is significantly higher than that in the 3.0 nm pores during the displacement process. This indicates a negative correlation between CH_4 desorption rate and slit pore width. This phenomenon can be explained as follows: as established in Sect. 2.4.1, the number of pre-adsorbed CH_4 molecules at equilibrium is consistent across different slit widths. Therefore, narrower pores contain a higher proportion of adsorbed CH_4 molecules. Once heteronuclear gas molecules begin to be injected, the concentration gradient between the adsorbed and free CH_4 molecules in the slit pores increases with decreasing pore width, leading to a faster desorption rate and a greater cumulative amount of desorbed CH_4. By the end of the displacement process, the cumulative desorption quantities of CH_4 in the 1.5 and 3.0 nm pores are 283 and 153 molecules,

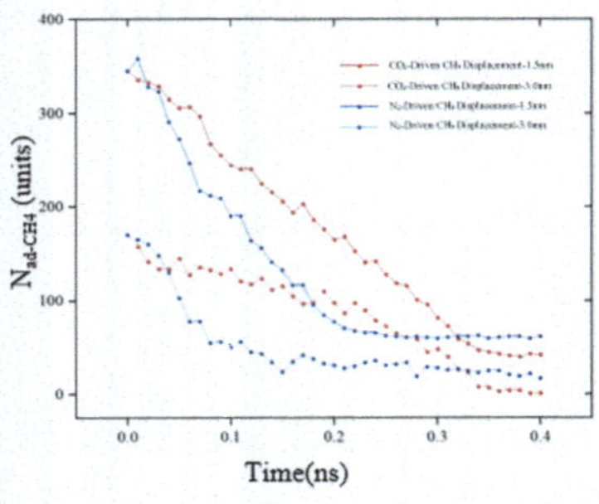

(a) Number of adsorbed CH_4 molecules

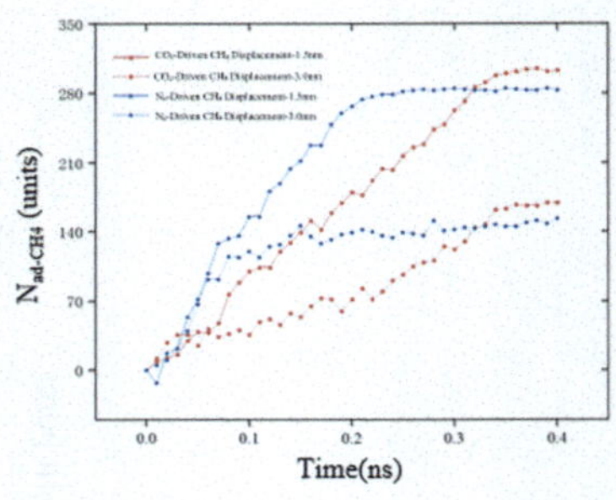

(b) Variation in CH_4 desorption amount

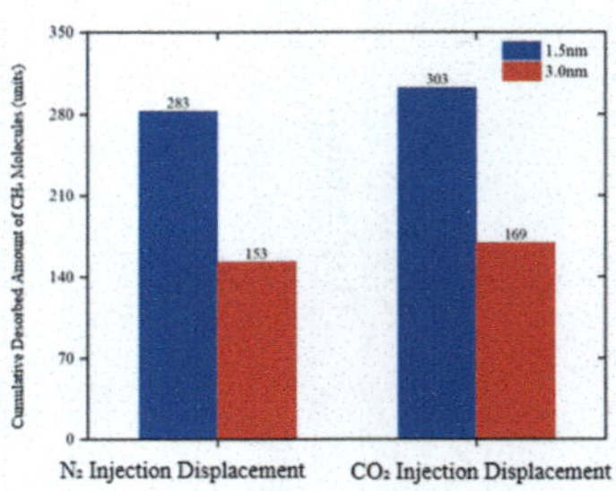

(c) Cumulative desorption amount of CH_4molecules

Fig. 2.19 Variation of the number of adsorbed-state CH_4 molecules during injection of CO_2/N_2 drive for each slit pore width class. (**a**) Number of adsorbed CH_4 molecules. (**b**) Variation in CH_4 desorption amount. (**c**) Cumulative desorption amount of CH_4 molecules respectively, under N_2 injection; under CO_2 injection, the corresponding values are 303 and 169 molecules.

Table 2.7 presents the residual quantity of adsorbed CH_4 molecules, total desorption amount, percentage of desorbed CH_4 relative to the initial adsorption, and the total time required for desorption at the end of the displacement process for different slit pore widths. As discussed in Sect. 2.4.1, narrower slit pores exhibit stronger adsorption strength toward CH_4 molecules, making it more difficult for CH_4 to detach from the pore walls. Consequently, at the end of the displacement process, the 1.5 nm slit pores retain more residual adsorbed CH_4 molecules compared to the 3.0 nm pores. Additionally, the percentage of desorbed CH_4 relative to the initially adsorbed amount is lower, and the time required to reach the end of the displacement process is longer in narrower pores.

2.5.2.2 Displacement Amount and Recovery Efficiency of CH_4 Molecules in Slit Pores

Figure 2.20 illustrates the variation in CH_4 displacement amount and recovery efficiency over time during the CO_2/N_2 injection-driven displacement process within slit pores of two different widths: 1.5 and 3.0 nm.

Table 2.7 Residual amount of CH_4 in adsorbed state, percentage of molecular weight of original adsorbed CH_4, and time for desorption

Slit width (nm)	Gas composition	Desorbed quantity with percentage contribution (个、%)	Residual adsorbed CH_4 quantity (个)	Desorption duration (ns)
1.5	N_2	283(82.03%)	62	0.23
	CO_2	303(87.83%)	42	0.38
3.0	N_2	153(90.00%)	17	0.16
	CO_2	169(99.41%)	1	0.34

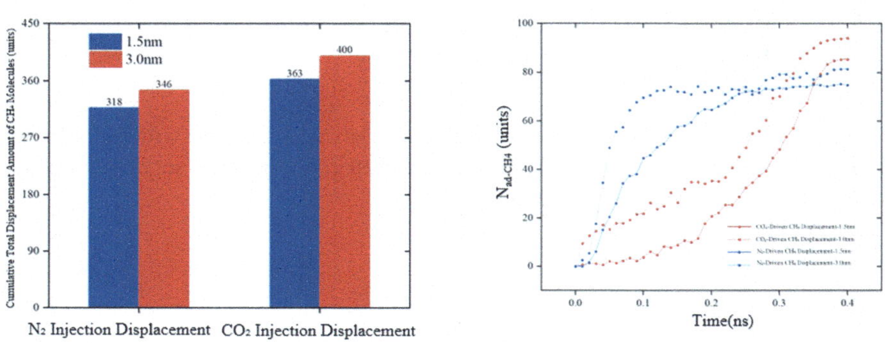

(a) Displacement molecular weight of CH_4 (b) CH_4 Molecular Recovery Rate

Fig. 2.20 Changes of CH_4 replacement and recovery in the slit pore during injection of CO_2/N_2 replacement at each slit pore width grade. (**a**) Displacement molecular weight of CH_4. (**b**) CH_4 molecular recovery rate

Table 2.8 CH_4 molecular drive volume and recovery in slit pore space

	1.5 nm		3.0 nm	
Temperature	Displacement	Recovery rate	Displacement	Recovery rate
CO_2 displacement	363	85.41%	400	94.12%
N_2 displacement	318	74.82%	346	81.41%

It can be observed that, regardless of whether N_2 or CO_2 is injected into the slit pore, the amount of CH_4 displacement increases with the enlargement of the pore width. When N_2 is used as the injection gas, increasing the slit pore width from 1.5 to 3.0 nm results in an increase of 45 CH_4 molecules being displaced. Similarly, when CO_2 is used, the CH_4 displacement increases by 54 molecules. This indicates that increasing the slit pore width enhances the displacement capacity of CH_4 within the coal matrix and improves the overall CH_4 recovery efficiency.

As shown in Fig. 2.20 and Table 2.8, regardless of whether N_2 or CO_2 is used as the injection gas, the displacement rate of CH_4 molecules increases more rapidly with greater slit pore widths. Consequently, the time required to reach displacement equilibrium is shorter, and the final CH_4 recovery rate is significantly higher. This phenomenon can be attributed to two primary factors. First, increasing the slit pore width reduces the adsorption strength of the coal macromolecular layers on CH_4 molecules, making it easier for CH_4 to desorb from the pore surfaces. Second, wider pores allow more injected heteronuclear gas molecules (CO_2 or N_2) to enter the pore per unit time, thereby enhancing competitive adsorption or partial pressure effects, and facilitating the displacement of a larger number of CH_4 molecules. Therefore, from a pore-scale perspective, in the contexts of coalbed methane or shale gas production and gas disaster mitigation, implementing engineering techniques to increase the number and width of coal fractures can significantly improve CH_4 recovery rates. This not only shortens the required operational time but also reduces overall time and economic costs.

2.5.3 *Effect of Injection Gas Type on the Dynamic Transport Characteristics of CH_4 Molecules*

This section analyzes the displacement characteristics of CH_4 gas within slit pores under different injection gas types. The simulation conditions are set at a temperature of 298.15 K and a slit pore width of 3 nm.

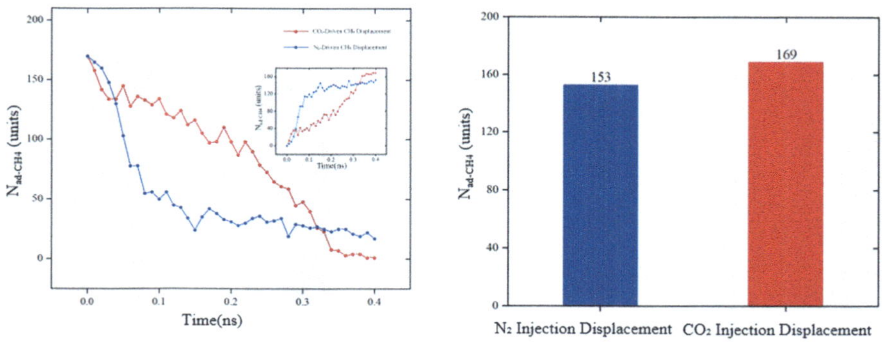

(a) Variation in Adsorbed CH_4 Quantity (b) Total Desorbed CH_4 Quantity

Fig. 2.21 Variation of CH_4 molecule desorption during CO_2/N_2 repulsion injection in slit pores. (**a**) Variation in adsorbed CH_4 quantity. (**b**) Total desorbed CH_4 quantity

2.5.3.1 The Variation in CH_4 Adsorption within the Slit Pores Is Examined under Different Injection Gas Types

As shown in Fig. 2.21a, during the early stage of the gas injection displacement process (0–0.03 ns), the desorption rate of adsorbed CH_4 molecules is significantly higher when CO_2 is injected compared to N_2. This is because, at the initial stage of injection, even a small number of CO_2 molecules entering the slit pore can readily compete with CH_4 molecules for adsorption sites on the pore wall due to CO_2's stronger adsorption affinity. This competitive adsorption effect effectively displaces CH_4 molecules from the adsorption sites, promoting their desorption. In contrast, since coal exhibits weaker adsorption capacity for N_2 than for CH_4, the limited number of N_2 molecules introduced during the early phase does not create sufficient partial pressure to trigger notable desorption, and the competitive adsorption effect is negligible. As gas injection continues (0.03–0.08 ns), the number of injected gas molecules inside the slit pore increases. During this stage, the desorption rate of CH_4 under N_2 injection surpasses that under CO_2 injection. This can be attributed to the pressure gradient generated by N_2 injection, which becomes sufficient to induce a rapid release of CH_4 within a short time. Meanwhile, the desorption rate under CO_2 injection remains relatively constant throughout the process, continuing until equilibrium is reached (as indicated by the stabilization of adsorbed CH_4 quantity within the slit). In the mid-to-late stage of gas injection (0.08–0.4 ns), the CH_4 desorption rate under N_2 injection slows down, whereas the desorption rate under CO_2 injection once again exceeds that of N_2. At this stage, the dominant mechanism driving further CH_4 desorption under CO_2 injection is competitive adsorption, as CO_2 molecules increasingly occupy surface sites, thereby displacing more CH_4 molecules.

Furthermore, as shown in Fig. 2.21b, CO_2 injection results in the desorption of more CH_4 molecules compared to N_2 injection. Under conditions of 298.15 K and a

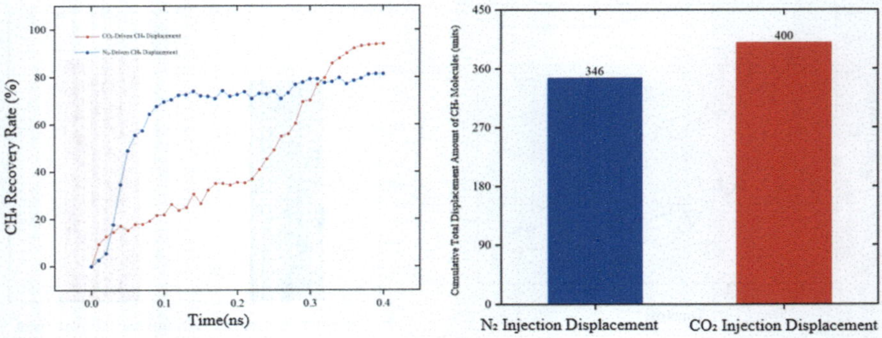

(a) CH_4 Molecular Recovery Rate (b) Displacement molecular weight of CH_4

Fig. 2.22 Changes in CH_4 drive volume and recovery rate during CO_2/N_2 drive injection in slit pores. (**a**) CH_4 molecular recovery rate. (**b**) Displacement molecular weight of CH_4

slit pore width of 3 nm, the number of CH_4 molecules desorbed during CO_2 injection exceeds that of N_2 injection by 16 molecules. This indicates that, from a pore-scale perspective, N_2 injection can more rapidly reduce the amount of adsorbed CH_4 in coal to a relatively high critical level, while CO_2 injection is more effective in achieving a more complete desorption of CH_4 from the coal matrix.

2.5.3.2 Variation in CH_4 Molecule Displacement and Recovery Efficiency within Slit Pores

Figure 2.22 shows the variation in CH_4 recovery and recovery efficiency over time during the CO_2/N_2 injection-driven displacement process in slit pores.

It can be observed that during the initial stage of displacement (0–0.03 ns), the rate of CH_4 recovery is higher under CO_2 injection compared to N_2 injection, a phenomenon previously explained. In the mid-stage of displacement (0.03–0.15 ns), the rate of CH_4 displacement increases more rapidly with N_2 injection than with CO_2. This is because a large number of N_2 molecules enter the slit pores, and their flow facilitates the removal of free CH_4 molecules through a "carrying" effect. The rapid reduction in free CH_4 leads to a concentration gradient between the adsorbed and free phases, promoting further desorption of CH_4 from the pore walls. In contrast, when CO_2 is injected, due to its stronger adsorption affinity for the coal matrix compared to CH_4, most CO_2 molecules preferentially adsorb onto the pore walls rather than transporting free CH_4 molecules. Although some free CH_4 is displaced by CO_2 flow, the overall growth rate of CH_4 displacement is slower during this stage. In both the initial and mid-stages, the displaced CH_4 primarily originates from the free phase. During the late stage of displacement (0.15–0.40 ns), the rate of CH_4 displacement under N_2 injection slows significantly, while under CO_2 injection, it accelerates. This is because competitive adsorption becomes the dominant mechanism in this phase. Since the coal matrix has a much stronger affinity for CO_2 than for N_2, the displacement capacity of N_2 is limited, resulting in a slower increase in CH_4 displacement in the back-pressure region. In contrast, CO_2 exhibits a stronger ability to displace adsorbed CH_4 through competitive adsorption. As the amount of CO_2 in the slit pores increases, this competition intensifies, leading to an accelerated rate of CH_4 displacement until the process reaches equilibrium. During this stage, most of the displaced CH_4 originates from the adsorbed phase on the pore walls.

Overall, under N_2 injection, the total number of CH_4 molecules displaced is 346, corresponding to a recovery rate of 81.41%. In comparison, CO_2 injection results in a total displacement of 400 CH_4 molecules, with a recovery rate of 94.12%. These results indicate that, from the perspective of microscopic slit pores, N_2 injection can displace the majority of CH_4 molecules in a relatively short time, while CO_2 injection is more effective at displacing the strongly adsorbed CH_4 molecules that are otherwise difficult to desorb. In other words, N_2 injection is advantageous for rapid early-stage displacement, whereas CO_2 injection excels in achieving a more complete final displacement.

2.6 Mechanism of CH_4 Molecular Transport and Production During Gas Injection Displacement at the Pore Scale

Figures 2.23 and 2.24 illustrate the variations in molecular number density of the injected gases CO_2 and N_2 within the slit pores over the course of the displacement process. The spatial distribution of gas molecule number density along the longitudinal direction of the slit pores is shown at specific time points: 0.01, 0.02, 0.03, 0.04, 0.1, 0.2, 0.3, and 0.4 ns. It can be observed that the transport behaviors of CO_2 and N_2 within the slit pores exhibit noticeable differences.

For CO_2 injection displacement, CO_2 molecules begin adsorbing onto the slit pore walls at the very early stages of the process. This is evidenced by the immediate formation of pronounced adsorption peaks on both sides of the slit pore, while the number of CO_2 molecules in the free region remains relatively low during the initial phase. As the displacement progresses, the number density of CO_2 molecules in the free region gradually increases, yet it consistently remains lower than the peak density observed in the adsorption zones. Additionally, as shown in the simulation snapshots of the CO_2 injection process in Fig. 2.23, the spatial distribution of CO_2 molecules along the longitudinal direction of the slit pore forms a characteristic "C"-shaped pattern. This indicates that CH_4 molecules originally adsorbed on the pore walls are desorbed due to the competitive adsorption effect of CO_2 molecules. In summary, during CO_2 injection, CO_2 molecules preferentially adsorb onto the slit pore walls and engage in competitive adsorption with CH_4 molecules. At the same time, there is also a "carrying" effect from the CO_2 flow that helps transport CH_4 molecules out of the slit pores.

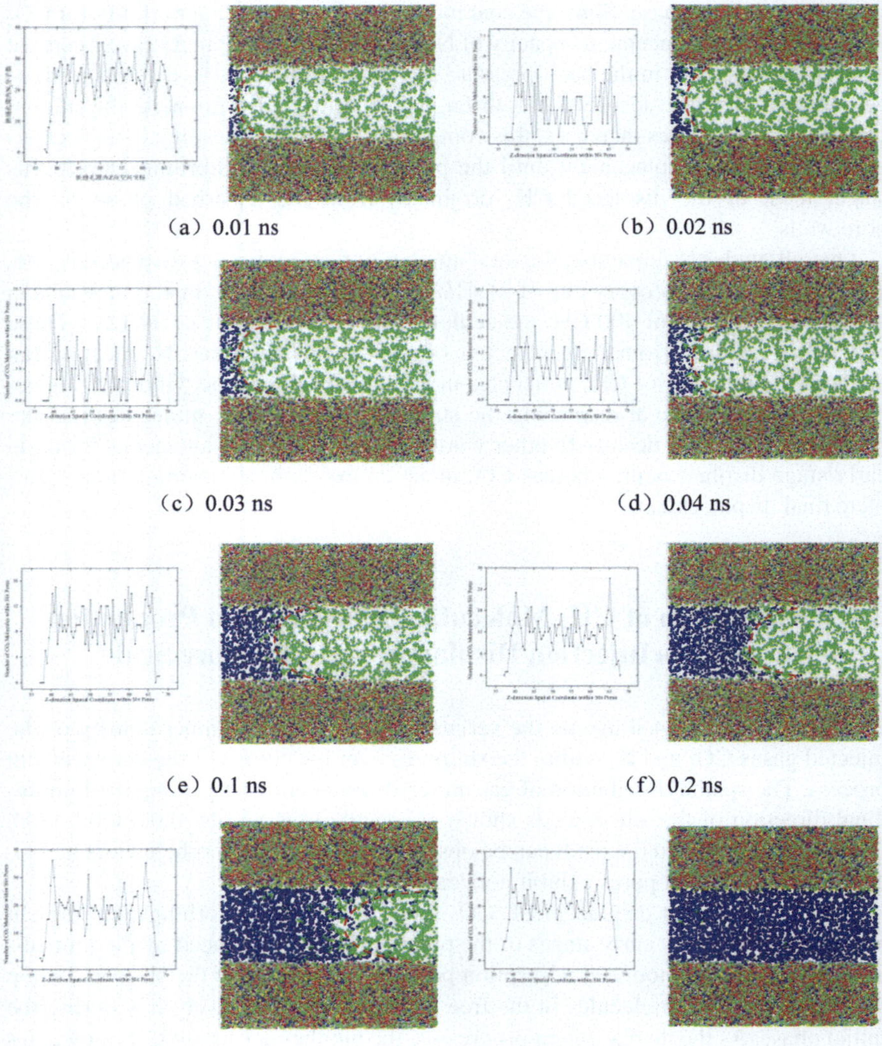

Fig. 2.23 Variation of molecular number density within the slit pores of the CO_2 injection repellent. (**a**) 0.01 ns. (**b**) 0.02 ns. (**c**) 0.03 ns. (**d**) 0.04 ns. (**e**) 0.1 ns. (**f**) 0.2 ns. (**g**) 0.3 ns. (**h**) 0.4 ns

As shown in Fig. 2.24, during N_2 injection displacement, the number density of N_2 molecules within the slit pores exhibits a distribution characterized by a higher concentration in the center and lower concentrations near the pore walls. This indicates that N_2 molecules predominantly enter the free region of the slit pore. Through the "carrying" effect of N_2 flow, free CH_4 molecules in the pore are displaced, which in turn increases the concentration gradient of CH_4 between the adsorption and free

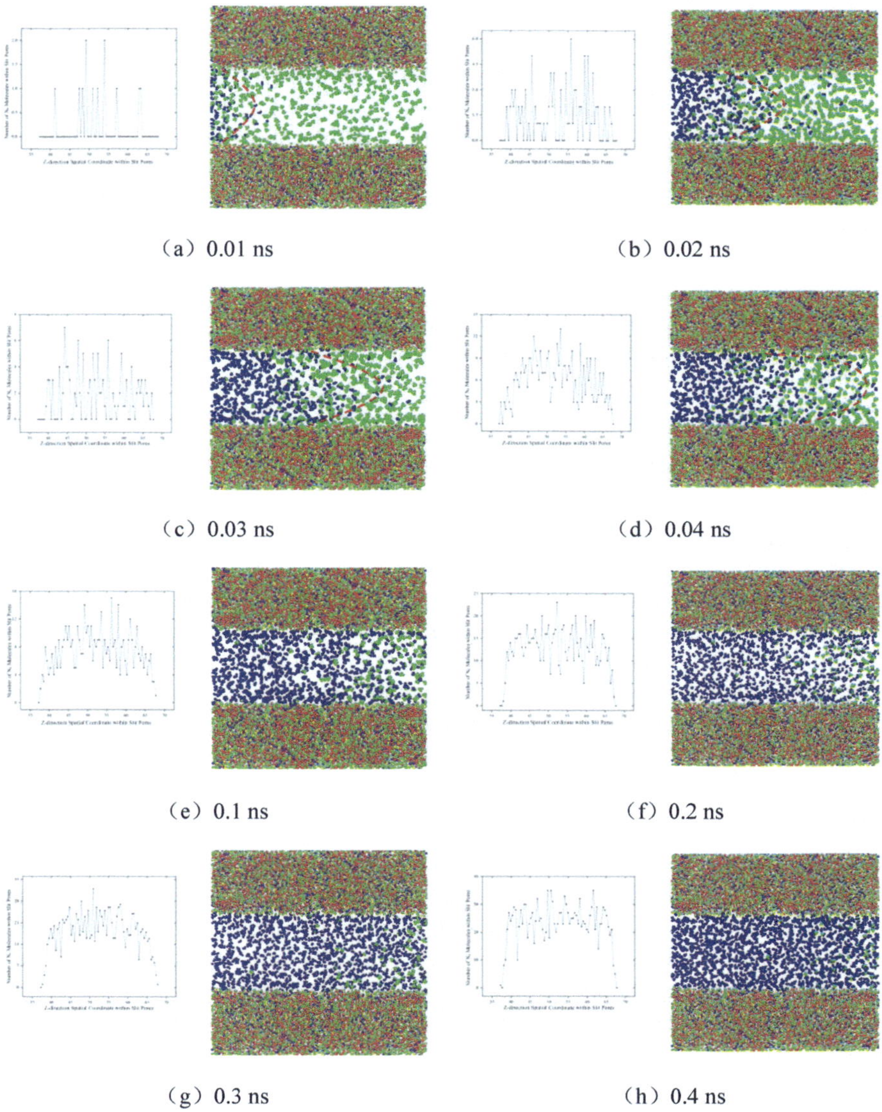

Fig. 2.24 Variation of molecular number density within the N_2 injection-driven slit pores. (**a**) 0.01 ns. (**b**) 0.02 ns. (**c**) 0.03 ns. (**d**) 0.04 ns. (**e**) 0.1 ns. (**f**) 0.2 ns. (**g**) 0.3 ns. (**h**) 0.4 ns

regions. This enhanced gradient promotes the desorption of CH_4 molecules originally adsorbed on the pore walls. From the simulation snapshots, it can be observed that in the early stages of the displacement process, the longitudinal distribution of N_2 molecules within the slit pore resembles an inverted "C" shape and that N_2 molecules quickly reach the outlet end of the slit pore. This further confirms that the N_2

injection displacement process primarily facilitates CH_4 desorption through a combination of the carrying effect and partial pressure gradients, thereby increasing the overall displacement of CH_4 molecules from the slit pores.

Overall, at the pore scale, the mechanisms by which CH_4 molecules are displaced within slit pores differ significantly between CO_2 and N_2 injection processes. During CO_2 injection, CO_2 molecules preferentially adsorb onto the slit pore walls and primarily reduce the amount of CH_4 in the pore through competitive adsorption with CH_4 and, to a lesser extent, via the carrying effect. In contrast, during N_2 injection, N_2 molecules initially act through a rapid carrying effect to remove free-phase CH_4 molecules from the slit pore. This process increases the concentration gradient of CH_4 between the adsorption and free regions, promoting desorption through partial pressure effects and thereby reducing the overall CH_4 content within the slit pores.

2.7 Chapter Summary

In this chapter, a microscopic gas displacement model of coal matrix slit pores capable of providing a continuous pressure gradient was established using MS and LAMMPS software. The model was used to analyze the adsorption, desorption, and displacement behavior of CH_4 molecules at the pore scale. The effects of temperature, slit pore width, and gas injection type on the dynamic transport behavior of CH_4 were systematically investigated. In addition, the dynamic transport characteristics and displacement fronts of hetero-gas molecules during the gas injection process were analyzed, thereby revealing the dynamic migration mechanisms of CH_4 molecules under displacement conditions at the microscopic pore scale. The specific conclusions are as follows:

1. Regarding the adsorption behavior of pure CH_4 gas molecules within slit pores, the results indicate that as the system temperature increases, both the density and peak value of CH_4 molecules in the adsorption region decrease, while the density in the free region increases. This suggests that elevated temperatures are unfavorable for the adsorption of CH_4 molecules onto the slit pore walls. Moreover, the width of the adsorption region remains consistent across all temperature levels. Additionally, as the slit pore width decreases, the adsorption capacity of the pore walls for CH_4 molecules increases, resulting in a stepwise increase in the number of adsorbed CH_4 molecules and a corresponding decrease in the number of free CH_4 molecules. This implies that the narrower the slit pores in the coal matrix, the greater the adsorption width along the pore walls.
2. Regarding the desorption characteristics of pure CH_4 gas molecules within slit pores, the desorption process exhibits a common trend across all temperature levels: the desorption amount of CH_4 initially increases rapidly, then gradually slows, and eventually reaches a stable state. As the system temperature rises, the rate of desorption of adsorbed CH_4 molecules accelerates, leading to a higher cumulative desorption and a greater final release of CH_4 molecules. Furthermore,

at a constant temperature, narrower slit pores result in both a faster initial desorption rate and a larger total desorption amount of CH_4. Conversely, wider slit pores yield a greater overall CH_4 release and a higher initial desorption rate.

3. Considering the influence of temperature on the dynamic transport behavior of CH_4 molecules during the gas injection displacement process at the microscale pore level: higher temperatures accelerate the desorption rate of CH_4 molecules, but lower temperatures result in a greater overall amount of desorbed CH_4. This indicates that, in the context of displacement operations, increasing temperature—whether injecting N_2 or CO_2—is not conducive to the desorption of adsorbed CH_4 within slit pores. Additionally, regardless of whether N_2 or CO_2 is used, rising temperatures lead to a faster initial increase in CH_4 displacement quantity, a shorter time required to reach displacement equilibrium, and ultimately greater CH_4 displacement volumes and recovery rates.
4. Considering the influence of slit pore width on the dynamic transport behavior of CH_4 molecules during gas injection displacement at the microscale: as the slit pore width decreases, regardless of whether N_2 or CO_2 is injected, the desorption rate of adsorbed CH_4 molecules within narrower pores is faster during the displacement process. However, by the end of displacement, a greater amount of adsorbed CH_4 remains in the narrower pores, the desorbed proportion relative to the initially adsorbed CH_4 is lower, and the time required to complete displacement is longer. Conversely, as the slit pore width increases, the rate of CH_4 displacement increases, and the total amount of CH_4 displaced from the pores is also higher.
5. Considering the influence of injection gas type on the dynamic transport behavior of CH_4 molecules during gas displacement at the microscale:N_2 injection can displace most of the CH_4 molecules from the slit pores in a short time, while CO_2 injection is more effective in displacing the adsorbed CH_4 molecules that are difficult to desorb. Based on the transport characteristics of N_2 and CO_2 molecules within slit pores, it is evident that the displacement mechanisms of CH_4 differ between CO_2 and N_2 at the microscale. Specifically, during N_2 injection, N_2 molecules primarily rely on a "carrying" effect and partial pressure gradient to quickly displace the majority of CH_4 molecules from the slit pores, with the displacement front forming a reverse "C" shape. In contrast, during CO_2 injection, CO_2 molecules preferentially adsorb onto the pore walls and displace CH_4 molecules more thoroughly through competitive adsorption and carrying effects, resulting in a "C"-shaped displacement front within the slit pores.

Chapter 3
LBM Simulation of Dynamic CH_4 Migration in Coal During Gas Injection and Displacement

3.1 Introduction

Coal is typically regarded as a heterogeneous porous medium with discrete pore and fracture structures [171]. Owing to this unique structure, the dynamic migration of CH_4 gas during gas injection and displacement in coal can be divided into two distinct components: (1) Free-phase CH_4 undergoes seepage through the coal's fracture network toward the production borehole/well, driven by a pressure gradient. (2) As injected gas increases and CH_4 flows out, the CH_4 content and concentration within the fractures gradually decrease. Under the CH_4 concentration gradient established between the coal matrix blocks and the fractures, adsorbed CH_4 within the matrix desorbs and diffuses into the fractures. At the pore-scale level, the migration of CH_4 gas molecules within the coal matrix pores during flooding primarily involves desorption and diffusion mechanisms. Scaling up the investigation, the dynamic migration of CH_4 gas during gas injection and displacement encompasses multiple transport behaviors—desorption, diffusion, and seepage—occurring within the complex pore–fracture system. Consequently, the dynamic migration behavior of CH_4 gas at this scale is fundamentally different from that observed at the microscopic pore level.

Due to the heterogeneity of coal-rock geometric morphology, variations in pore distribution, and multiple migration mechanisms of fluids, both experimental and theoretical studies on the dynamic migration behavior of CH_4 gas during displacement processes face significant challenges. In recent years, researchers have proposed various numerical simulation methods. One approach is based on the principle that macroscopic systems consist of numerous microscopic particles. By tracking and statistically accumulating the motion of each microscopic particle governed by mechanical laws, this method derives macroscopic fluid behavior. However, it requires enormous computational resources and places high demands on computer processing power and storage capacity. An alternative method begins by

G. Wang, H. Gong, *Dynamic Migration Characteristics of CH4 During Gas Injection Displacement in Coal Seams*,
https://doi.org/10.1007/978-981-95-8879-4_3

establishing unified rules for microscopic particles and then explores macroscopic physical phenomena. The lattice Boltzmann method (LBM), a computational fluid dynamics approach, follows this paradigm. As a discretized hydrodynamic numerical simulation method, LBM utilizes microscopic kinetic behaviors of fluids in discrete velocity spaces and lattice systems to simulate macroscopic fluid dynamics. The LBM divides computational domains into lattice grids while discretizing spatial, temporal, and velocity parameters of fluid flow. It abstracts fluid motion as collision and migration processes of microscopic particles (fluid micro-clusters) across lattice nodes, strictly adhering to the conservation laws of mass, momentum, and energy. Functioning as a continuous method at macroscopic scales while maintaining discrete characteristics at microscopic levels, LBM serves as a bridge connecting macro-micro and continuum-discrete perspectives, demonstrating exceptional capability in modeling complex fluid flow phenomena.

This chapter investigates the dynamic migration behavior of CH_4 gas in coal during gas injection displacement using the LBM. A coupled LBM model integrating the Navier-Stokes equations (NSE) and advection-dispersion equations (ADE) is first developed to account for the directional flow of mixed gases and mass transfer processes inside and outside the coal matrix. Building on this model, the dynamic migration patterns of CH_4 during gas injection displacement are simulated, with specific focus on the effects of injection flow rate and types of injectant gases on CH_4 migration behavior. Based on these findings, the dynamic migration-production mechanisms of CH_4 in coal during gas injection displacement are comprehensively analyzed.

3.2 Fundamental Theory of the LBM Method in Gas Injection and Displacement Processes

3.2.1 Core Model of LBM

LBM is a simulation approach grounded in the kinetic theory of gases, having evolved from the lattice gas automata (LGA) method [172]. It constitutes a discrete scheme for solving the continuous Boltzmann equation. The Boltzmann equation can be expressed by Eq. (3.1):

$$\frac{\partial f(x,t)}{\partial t}+\xi\frac{\partial f(x,t)}{\partial r}+a\frac{\partial f(x,t)}{\partial \xi}=\Omega(f) \tag{3.1}$$

where $f(x,t)$ denotes the particle distribution function at position x within the infinitesimal range $[x, x + dx]$ and velocity ξ within $[\boldsymbol{\xi}, \boldsymbol{\xi} + d\boldsymbol{\xi}]$ at time t, a represents the acceleration due to external forces, ξ is the particle velocity vector, and $\Omega(f)$ denotes the collision term. The simplified form of $\Omega(f)$ is given by Eq. (3.2):

$$\Omega(f) = -\frac{1}{\tau}\left[f(x,t) - f^{eq}(x,t)\right] \tag{3.2}$$

In the equation, τ is the dimensionless relaxation time, and f^{eq} is the equilibrium distribution function.

LBM models can be classified into the single relaxation time model (SRT) and the multiple relaxation time model (MRT) based on the number of relaxation times included in the collision term.

3.2.1.1 Single Relaxation Time Model (SRT)

The single relaxation time model generally refers to the BGK (Bhatnagar–Gross–Krook) collision model [173]. As a key implementation approach in LBM, this model represents a widely used simplified collision operator and stands as one of the most extensively applied variants. It employs a single relaxation time parameter to describe the process of particle distribution functions evolving toward equilibrium. The basic form of the collision operator can be expressed as shown in Eq. (3.3):

$$f_i(x + c_i\Delta t, t + \Delta t) - f_i(x,t) = -\frac{1}{\tau}\left[f_i(x,t) - f_i^{eq}(x,t)\right] \tag{3.3}$$

In the equation, $f_i(x,t)$ represents the distribution function along the discrete velocity direction c_i at position x and time t; c_i denotes the discrete velocity of particles; Δt is the time step; $f_i^{eq}(x,t)$ is the local equilibrium distribution function, whose form is typically determined by macroscopic fluid properties (such as density and velocity); and τ is the relaxation time, related to the dynamic viscosity ν of the fluid, with their relationship given by Eq. (3.4):

$$\nu = c_s^2(\tau - 0.5)\Delta t \tag{3.4}$$

In the equation, c_s is the lattice sound speed.

Equation (3.3) can be decomposed into two processes: collision and streaming:

Collision Process

In the single relaxation time model, the collision process simulates the evolution of particle distribution functions toward equilibrium through a simplified mechanism. Governed by a single relaxation time parameter τ, this process describes how the distribution function f_i exponentially approaches its local equilibrium state f_i^{eq}. The collision operator is given by Eq. (3.5):

$$f_i^*(x,t) = f_i(x,t) - \frac{1}{\tau}\left[f_i(x,t) - f_i^{eq}(x,t)\right] \tag{3.5}$$

where f_i is the post-collision distribution function.

Streaming Process

Following the collision step, the particle distribution functions undergo streaming, propagating along their respective discrete velocity directions to adjacent lattice sites. This process is described by Eq. (3.6):

$$f_i(x + c_i\Delta t, t + \Delta t) = f_i^*(x,t) \tag{3.6}$$

Following the streaming step, the updated distribution functions at new lattice sites proceed to the next collision-streaming cycle. Currently, the most widely adopted equilibrium distribution function and discrete velocity model is the DnQb framework proposed by Qian et al. [174], where n denotes spatial dimensionality and b represents the number of discrete velocities. Its equilibrium distribution function takes the form shown in Eq. (3.7):

$$f_i^{eq}(x,t) = \rho\omega_i\left[1 + \frac{c_i u}{c_s^2} + \frac{uu\left(c_i^2 - c_i^2 I\right)}{2c_s^4}\right] \tag{3.7}$$

where I denotes the identity tensor; ρ represents particle density; u is the particle velocity; and ω stands for the weighting coefficient. For simulating gas transport behavior in two and three dimensions, commonly used lattice models include D2Q5, D2Q9, D3Q15, and D3Q19. Detailed descriptions and parameter specifications for these models are comprehensively provided in Ref. [170] and will not be reiterated here.

Following the computation of microscopic particle states, the macroscopic fluid properties are calculated using Eqs. (3.8), (3.9), and (3.10), thereby achieving the transformation from microscopic to macroscopic quantities:

$$\text{Macroscopic density}: \rho = \sum_i f_i \tag{3.8}$$

$$\text{Macroscopic density}: \rho u = \sum_i c_i f_i \tag{3.9}$$

$$\text{Macroscopic pressure}: p = c_s^2 \rho \tag{3.10}$$

3.2.1.2 Multiple Relaxation Time (MRT) Model

The multiple relaxation time (MRT) model [175] enhances simulation stability and accuracy by introducing distinct relaxation times for different modes of particle distribution functions. This approach is grounded in the concept that fluid distribution functions relax independently across various kinetic modes. The collision term is defined through the collision matrix $-\Lambda$, with the collision operator $\Omega(f)$ defined in Eq. (3.11) as:

$$\Omega_i(f) = -\Lambda_{ij}\left(f_j - f_j^{eq}\right) \tag{3.11}$$

Similar to the single relaxation Boltzmann equation, the multiple relaxation time Boltzmann equation can also be decomposed into collision and streaming processes, as expressed in Eqs. (3.12) and (3.13):

$$\text{Collision process}: f_i^*(x,t) = f_i(x,t) - \Lambda_{ij}\left(f_j - f_j^{eq}\right) \tag{3.12}$$

$$\text{Streaming process}\, f_i\left(x + c_i\Delta t, t + \Delta t\right) = f_i^*(x,t) \tag{3.13}$$

By multiplying both sides of Eq. (3.12) by the transformation matrix M, we derive the collision process in moment space, as shown in Eq. (3.14):

$$M^* = m - S\left(m - m^{eq}\right) \tag{3.14}$$

In the equation, m^{eq} denotes the equilibrium distribution function in moment space, and S represents the relaxation matrix, typically a diagonal matrix defined by Eq. (3.15):

$$S = M\Lambda M^{-1} = \begin{bmatrix} s_1 & & \\ & \cdots & \\ & & s_{ij} \end{bmatrix} \tag{3.15}$$

The above formulation demonstrates that different moment functions m_i can be independently controlled through distinct relaxation times $1/s_i$.

While the evolution of the multiple relaxation time lattice Boltzmann model can also be constructed on various discrete velocity models across dimensions, Qian et al. [176] provide a comprehensive summary of relevant parameters for the MRT collision model (using the D2Q9 framework as an example), which will not be repeated here.

The single relaxation time model requires only one relaxation parameter to be determined, minimizes computational overhead, and suffices for the research objectives of this chapter. Therefore, the single relaxation time model has been selected as the collision term in the lattice Boltzmann model developed in subsequent sections.

3.2.2 LBM Boundary Treatment Schemes

In the LBM framework, the implementation of boundary conditions is pivotal for ensuring numerical accuracy and stability. Unlike macroscopic methods that directly apply boundary conditions using macroscopic quantities (pressure, velocity, etc.), the LBM requires determination of particle distribution functions based on prescribed macroscopic boundary values. The boundary schemes employed in this chapter include: periodic boundary format [177], bounce-back boundary format [178], kinetic-based boundary format [179], extrapolation boundary format [180].

3.2.2.1 Periodic Boundary Scheme

Periodic boundary condition represents a widely employed boundary treatment, as illustrated in Fig. 3.1. They are particularly well-suited for simulating flow fields with repeating structures or infinite-domain flow problems. The core concept assumes physical continuity of fluid properties between paired opposing boundaries: when fluid particles exit one boundary, they immediately re-enter the domain through the opposite boundary with identical physical attributes. This approach effectively models either unbounded flows or confined flows within a repetitive unit.

Consider a two-dimensional computational domain with dimensions L_x (x-direction) and L_y (y-direction). The mathematical formulation of periodic boundary conditions is given by Eqs. (3.16) and (3.17):

$$f_i\left(x+L_x,y,t\right)=f_i\left(x,y,t\right) \tag{3.16}$$

$$f_i\left(x,y+L_y,t\right)=f_i\left(x,y,t\right) \tag{3.17}$$

Fig. 3.1 Schematic of cycle boundary format

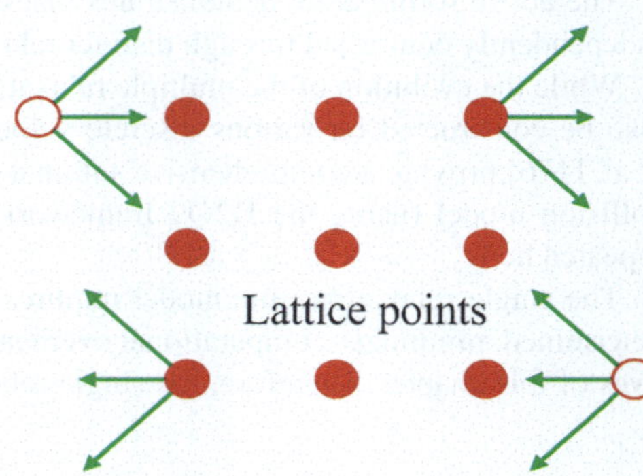

In the equations, f_i denotes the distribution function along the ith discrete velocity direction, and x and y represent spatial coordinates within the computational domain.

3.2.2.2 Bounce-Back Boundary Scheme

The bounce-back boundary condition primarily simulates no-slip conditions at solid walls. It assumes fluid particles striking a solid boundary reverse their velocity vectors upon impact, as illustrated in Fig. 3.2. This mechanism enforces zero velocity at solid boundaries, thereby modeling the no-slip condition. Through this approach, the bounce-back scheme effectively captures fluid–solid interactions.

In LBM implementation, the bounce-back boundary modifies the evolution of particle distribution functions f_i. Consider a solid wall within the computational domain: when fluid particles collide with the wall, their distribution functions f_i are redirected to the opposite direction i^*. The mathematical formulation is given by Eq. (3.18):

$$f_i\left(x_b, t+\Delta t\right) = f_{i^*}\left(x_b, t\right) \tag{3.18}$$

In the equation, x_b denotes the boundary node coordinate (solid wall position), f_i denotes the distribution function along the ii-th direction, and i^* denotes the direction opposite to i.

The bounce-back scheme described above is referred to as the standard bounce-back format. While it strictly enforces mass and momentum conservation within the system, its spatial accuracy is limited to first-order. To ensure higher precision, improved formulations such as the second-order accurate half-way bounce-back scheme have been proposed [181, 182].

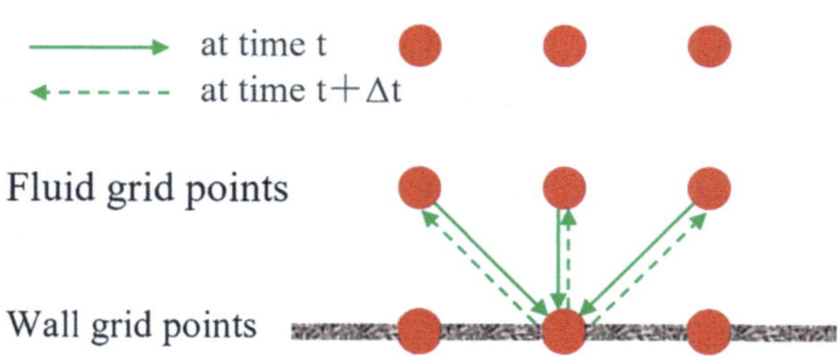

Fig. 3.2 Schematic of the bounce boundary format

3.2.2.3 Kinetic Boundary Scheme

The kinetic boundary condition is an effective approach for handling fluid boundaries, particularly suited for complex geometries and dynamically evolving interfaces. Its fundamental principle relies on kinetic theory to directly modify distribution functions at boundary nodes to satisfy physical constraints. Unlike traditional macroscopic interpolation methods, this scheme leverages the properties of microscopic distribution functions to accurately simulate fluid behavior at boundaries through direct manipulation of distribution functions.

For isothermal models, the governing equation system is given by Eqs. (3.19) and (3.20):

$$\rho = \sum_i f_i \tag{3.19}$$

$$\rho u = \sum_i c_i f_i \tag{3.20}$$

In the equation, ρ denotes the density at the boundary node, and u denotes the velocity at the boundary node. Kinetic boundary schemes have evolved into various forms, among which the Zou-He non-equilibrium bounce-back Scheme [183] is the most widely adopted. This boundary treatment approach assumes that the bounce-back format is applied exclusively in the direction normal to the boundary, where unknown distribution functions in the normal direction are determined via the bounce-back mechanism, while unknown distribution functions in other directions are resolved by solving systems of equations based on macroscopic physical quantities at the boundary.

3.2.2.4 Extrapolation Boundary Scheme

The extrapolation boundary condition ensures that fluid physical quantities at boundaries conform to physical reality by extrapolating distribution functions. Rooted in hydrodynamic extrapolation principles, this method reconstructs fluid states at boundaries using flow field data from interior nodes. Specifically, it computes boundary distribution functions by extrapolating information from nearby internal nodes, enabling effective boundary condition treatment. This approach is particularly suited for simulating fluid boundaries with complex geometries and dynamic evolution.

The primary extrapolation formats include the non-equilibrium extrapolation Scheme [184] and the distribution function extrapolation Scheme [185]. The non-equilibrium scheme extrapolates boundary distribution functions by modifying distribution functions at interior nodes based on non-equilibrium components. This method typically relies on hydrodynamic properties near boundaries—such as density and velocity—for extrapolation. Conversely, the distribution function scheme directly extrapolates boundary distribution functions from internal nodes without relying on macroscopic quantities at boundaries; instead, it calculates distribution functions at boundary nodes purely through interpolation or extrapolation techniques.

3.2.3 LBM Dimensional Conversion

Due to the discrete nature of the lattice Boltzmann method, conventional units for physical quantities become incompatible. To facilitate computation and data processing, physical quantities are typically expressed in lattice units and converted to actual physical units through dimensionless scaling.

The conversion between actual physical units and lattice units proceeds as follows:

(a) The characteristic length L_0 is determined as follows: For a two-dimensional lattice field composed of $N \times N$ grid points, with a corresponding physical domain size of $L \times L$, the characteristic length is defined as $L_0 = lr = L/N$, where l_r represents the physical side length per lattice unit.

(b) The characteristic time T_0 is established through three sequential steps: First, determine the relaxation time τ; second, calculate the physical dynamic viscosity ν using Eq. (3.4); finally, derive T_0 via Eq. (3.21).

$$\nu = \nu_{phy} \frac{T_0}{L_0^2} \tag{3.21}$$

In the equation, ν_{phy} denotes the physical dynamic viscosity with units of m^2/s.

(c) The characteristic mass M_0 is determined as follows: Given the physical fluid density ρ_{phy} and the lattice density ρ, M_0 is derived using Eq. (3.22):

$$\rho = \rho_{phy} \frac{L_0^3}{M_0} \tag{3.22}$$

(d) Once the characteristic scales L_0 (length), T_0 (time), and M_0 (mass) are established, all other physical quantities can be expressed in dimensionless form as summarized in Table 3.1:

3.3 LBM Simulation of CH_4 Dynamic Transport During Gas Injection Displacement of Coalbed Methane

3.3.1 Governing Equations

Within LBM, the coupled Navier-Stokes equations (NSE) and advection-diffusion equation (ADE) are frequently employed to investigate single-component gas migration and gas-solid adsorption in two-dimensional porous media. However, this

Table 3.1 Conversion of lattice units to actual physical units

Physical quantity	Lattice unit symbol	Physical unit symbol	Conversion formula
Length	l	L (m)	$l = L/L_0$
Time	t	T (s)	$t = T/T_0$
Lattice spacing	Δx (generally taken as 1)	lr (m)	$\Delta x = l_r/L_0$
Time step	Δt (generally taken as 1)	ΔT (s)	$\Delta t = \Delta T/T_0$
Lattice speed	$c = \Delta x/\Delta t$(generally taken as 1)	C (m/s)	$c = C{\cdot}T_0/L_0$
Density	ρ	ρ_0 (kg/m^3)	$\rho = \rho_0/M_0$
Kinematic viscosity	v	v_0 (m^2/s)	$v = v_0{\cdot}T_0/L_0^2$
Velocity	u	U_0 (m/s)	$u = U_0{\cdot}T_0/L_0$
Body force	f	F (kg·m/s^2)	$f = F{\cdot}T_0^2/(M_0L_0)$
Pressure	p	P (kg/(m·s^2))	$p = P{\cdot}L_0T_0^2/M_0$

approach has seen limited application to multiphase and multicomponent flow. This chapter develops a flow-coupled LBM model based on the NSE-ADE [186] framework, incorporating multicomponent gas flow and mass transfer processes. This model is used to simulate the process of injecting gas to displace CH_4 from coal, aiming to study the dynamic transport behavior of heterocomponent gases and CH_4.

3.3.1.1 Navier-Stokes Governing Equations for Fluid Flow

The multicomponent gas mixture is driven by an external flow field, which is simulated as a fictitious carrier fluid with constant mass and momentum conservation. The multicomponent fluid flow is governed by the Navier-Stokes equations, as given in Eqs. (3.23) and (3.24):

$$\partial t\rho + \nabla \cdot (\rho u) = 0 \tag{3.23}$$

$$\partial t(\rho u) + \nabla \cdot (\rho uu) = -\nabla p + \nabla \cdot \sigma \tag{3.24}$$

where u denotes the velocity of the multicomponent fluid; p denotes the pressure of the multicomponent fluid; ρ denotes the density of the multicomponent fluid; and σ denotes the viscous stress tensor. The flow field is discretized into a lattice grid, and the D2Q9 model is employed to solve for the set of 9 discrete velocity vectors c_i at the lattice nodes, as expressed by Eq. (3.25):

$$c_i = \begin{cases} (0,0), & i = 0 \\ (\cos\theta_i, \sin\theta_i)c, & i = 1,2,3,4 \\ \sqrt{2}(\cos\theta_i, \sin\theta_i)c, & i = 5,6,7,8 \end{cases} \tag{3.25}$$

where $\theta_i = (i - 1)\pi/2$, and c denotes the lattice speed, taking the value $c = 1$.

The single relaxation time model (BGK model) is employed. The evolution equation for the particle distribution function f_i at position x over time t is given by Eqs. (3.26) and (3.27):

$$f_i\left(x+c_i\Delta t,t+\Delta t\right)-f_i\left(x,t\right)=-\frac{1}{\tau}\left[f_i\left(x,t\right)-f_i^{eq}\left(x,t\right)\right] \tag{3.26}$$

$$f_i^{eq}\left(x,t\right)=\rho\omega_i\left[1+\frac{c_i u}{c_s^2}+\frac{\left(c_i\cdot u\right)^2}{2c_s^4}-\frac{u^2}{2c_s^2}\right] \tag{3.27}$$

The definitions of all parameters in these equations are detailed in Sect. 3.2.1. Furthermore, within the D2Q9 lattice model, the lattice speed of sound c_s is assigned the value $3^{-1/2}$. The directional weight coefficients ω_i for directions 0 to 8 are given by Eq. (3.28):

$$\omega_i=\begin{cases}4/9 & i=0\\ 1/9 & i=1,2,3,4\\ 1/36 & i=5,6,7,8\end{cases} \tag{3.28}$$

Using Eqs. (3.29) and (3.30), the fluid density and velocity at each lattice node are obtained by summing the particle distribution functions over all discrete velocity directions:

$$\rho=\sum_i f_i \tag{3.29}$$

$$\rho u=\sum_i c_i f_i \tag{3.30}$$

The fluid pressure p is calculated by Eq. (3.31):

$$\rho=\sum_i f_i \tag{3.31}$$

3.3.1.2 Governing Equations for Gas Mass Transfer Within Coal Matrices and Across Coal Matrix Surfaces

The mass transfer processes of heterocomponent gases and CH_4 from the free space to the surface of the coal matrix are represented by two independent advection-diffusion equations, Eqs. (3.32) and (3.33):

$$\partial t C_{CO_2/N_2} + \nabla\left(C_{CO_2/N_2} u\right) = \nabla \cdot \left(D_{CO_2/N_2} \nabla C_{CO_2/N_2}\right) + R_{CO_2/N_2} \tag{3.32}$$

$$\partial t C_{CH_4} + \nabla\left(C_{CH_4} u\right) = \nabla \cdot \left(D_{CH_4} \nabla C_{CH_4}\right) + R_{CH_4} \tag{3.33}$$

where $C_{CO2/N2}$ and C_{CH4} denote the concentrations of the injected heterocomponent gas (COmponent gas (COons of the inject$D_{CO2/N2}$ and D_{CH4} denote the bulk diffusion coefficients of the heterocomponent gas (COmponent gas (COion coefficients of the heterocomponent gas ($R_{CO2/N2}$ and R_{CH4} denote the source/sink terms for the heterocomponent gas (COmponent gas (CO terms for the heterocomponente nodes, respectively, modeling gas adsorption/desorption phenomena on the coal surface. These terms are set to zero in the pore free space.

Employing the monolayer assumption of Langmuir adsorption kinetics, the source/sink term $R_{CO2/N2}$ for the injected heterocomponent gas (CO_2 or N_2) is given by Eq. (3.34):

$$R_{CO_2/N_2} = \frac{\partial V_{CO_2/N_2}}{\partial t} = k_{aCO_2/N_2} C_{CO_2/N_2} \left(V_{mCO_2/N_2} - V_{CO_2/N_2}\right) - k_{dCO_2/N_2} V_{CO_2/N_2} \tag{3.34}$$

where $V_{mCO2/N2}$ denotes the monolayer adsorption capacity of the injected heterocomponent gas (COmponent g $V_{CO2/N2}$ represents the adsorbed amount of the heterocomponent gas at time t; and $k_{aCO2/N2}$ and $k_{dCO2/N2}$ are the adsorption rate constant and desorption rate constant for the heterocomponent gas, respectively.

During actual gas injection displacement, the coal matrix surface is saturated with adsorbed CH_4, with its monolayer capacity denoted as V_{mCH4}. Upon injection of the heterocomponent gas (CO_2 or N_2), competitive adsorption due to partial pressure changes induces CH_4 desorption, reducing its effective adsorption capacity to V^*_{mCH4}. For this dynamic adsorption-desorption process, the source/sink term R_{CH4} at gas-solid boundary nodes is governed by Eq. (3.35):

$$\begin{cases} R_{CH_4} = \dfrac{\partial V_{CH_4}}{\partial t} = -k_{dCH_4} V_{CH_4} & if \quad V_{CH_4} > V^*_{mCH_4} \\ R_{CH_4} = \dfrac{\partial V_{CH_4}}{\partial t} = k_{aCH_4} C_{CH_4} \left(V^*_{mCH_4} - V_{CH_4}\right) - k_{dCH_4} V_{CH_4} & if \quad V_{CH_4} \le V^*_{mCH_4} \end{cases} \tag{3.35}$$

where V_{mCH4} denotes the monolayer adsorption capacity of CH, V_{CH4} denotes the adsorbed amount of CH capacity t; and k_{aCH4} and k_{dCH4} are the adsorption rate constant and desorption rate constant for CHrmes induces CH_4

Langmuir adsorption kinetics governs mass exchange across phases at the gas-solid boundary. Adsorbed gases subsequently penetrate the coal matrix interior via pure diffusion (with zero advective velocity), driven by concentration gradients from the boundary. The transport of adsorbed gases within the coal matrix is described by the homogeneous solid diffusion model (HDSM) [187], as given in Eqs. (3.36) and (3.37):

$$\frac{\partial V_{CO_2/N_2}}{\partial t} = D_{sCO_2/N_2}\left(\frac{\partial^2 V_{CO_2/N_2}}{\partial x^2} + \frac{\partial^2 V_{CO_2/N_2}}{\partial y^2}\right) \tag{3.36}$$

$$\frac{\partial V_{CH_4}}{\partial t} = D_{sCH_4}\left(\frac{\partial^2 V_{CH_4}}{\partial x^2} + \frac{\partial^2 V_{CH_4}}{\partial y^2}\right) \tag{3.37}$$

where $D_{SCO2/N2}$ and D_{SCH4} denote the diffusion coefficients of the heterocomponent gas (COmponent gas (COefficients of the heterocomponent gas (

The D2Q5 lattice model is employed to solve for the set of discrete concentration distribution functions C_i at the lattice nodes across 5 discrete velocity directions, as defined by Eq. (3.38):

$$C_i = \begin{cases} (0,0), & i = 0 \\ (\cos\theta_i, \sin\theta_i)c, & i = 1,2,3,4 \end{cases} \tag{3.38}$$

where $\theta_i = (i - 1)\pi/2$. The evolution equation for the particle distribution function φ_i over a single time step Δt is given by Eqs. (3.39) and (3.40):

$$\phi_i\left(x + C_i\Delta t, t + \Delta t\right) - \phi_i\left(x,t\right) = -\frac{1}{\tau^*}\left[\phi_i\left(x,t\right) - \phi_i^{eq}\left(x,t\right)\right] \tag{3.39}$$

$$\phi_i^{eq}\left(x,t\right) = C\omega_i^*\left[1 + \frac{C_i u}{c_s^2}\right] \tag{3.40}$$

where τ^* is the relaxation time for advection-diffusion; and the directional weight coefficients ω^*_i for directions 0 to 4 are given by Eq. (3.41):

$$\omega_i^* = \begin{cases} 1/3 & i = 0 \\ 1/6 & i = 1,2,3,4 \end{cases} \tag{3.41}$$

The gas diffusion coefficient D is related to the relaxation time τ^* through Eq. (3.42):

$$D = c_s^2\left(\tau^* - 0.5\right)\Delta t \tag{3.42}$$

The gas concentration C at each lattice node is obtained by summing the particle distribution functions over discrete velocity directions 0 to 4 using Eq. (3.41):

$$C = \sum_i \phi_i \tag{3.43}$$

A discrete set of source/sink terms q_i is incorporated into Eq. (3.39) to account for mass transfer across the coal matrix gas–solid interface [188], as expressed in Eqs. (3.44)–(3.46):

$$\phi_i\left(x+C_i\Delta t,t+\Delta t\right)-\phi_i\left(x,t\right)=-\frac{1}{\tau^*}\left[\phi_i\left(x,t\right)-\phi_i^{eq}\left(x,t\right)\right]+q_i\Delta t \tag{3.44}$$

$$q_i=\left(1-\frac{1}{2\tau^*}\right)\omega_i^* R \tag{3.45}$$

$$C=\sum_i \phi_i+\frac{1}{2}R\Delta t \tag{3.46}$$

3.3.2 *Model Validation*

The coupled flow-diffusion-adsorption process of gases within the porous coal structure is illustrated in Fig. 3.3. Here, coal matrix blocks are conceptualized as discrete particles. Larger-diameter interparticle pores exist between these coal matrix particles, while smaller-diameter intraparticle pores exist within the coal matrix itself. This characteristic bimodal pore size distribution leads to the division of gas transport in coal into two distinct processes:

(a) Between Coal Matrix Blocks: Gases migrate to the surfaces of coal matrix particles primarily via advection-diffusion through the interparticle pore network. Subsequently, they become adsorbed onto the particle surfaces through surface adsorption phenomena.
(b) Within the Coal Matrix: Due to the extremely small scale of the intraparticle pores, gas flow (advection) within these internal pores is negligible and thus disregarded. Consequently, gas transport inside the coal matrix occurs solely

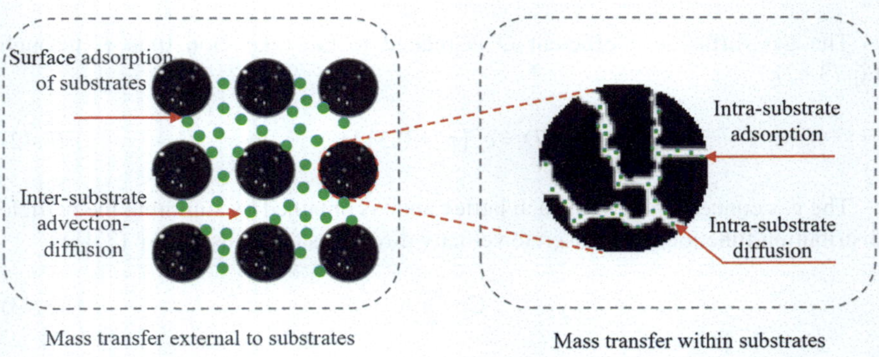

Fig. 3.3 Schematic diagram of gas transportation in a porous medium

via pure diffusion. Gases initially adsorbed on the particle surface diffuse further inward driven by concentration gradients, ultimately adsorbing onto the internal pore surfaces of the coal matrix.

Based on the two processes described above, two benchmark cases are established to validate the accuracy of the coupled NSE-ADE LBM model.

Case 1 Validation of Interparticle Gas Flow Process: The Poiseuille flow [189] accurately describes gas flow through the larger interparticle pores between coal matrix blocks. Its analytical solution is given by Eq. (3.47). For this validation, a single-channel model with dimensions $L_x \times L_y = 10 \times 100$ is simulated using the D2Q9 lattice scheme. The extrapolation boundary scheme is applied at the inlet/outlet (x-direction boundaries). To ensure the flow remains within the laminar flow regime, a small pressure gradient is imposed across the channel.

$$u_{\max} = \frac{\nabla P_x}{2\mu} y(b-y) \tag{3.47}$$

where u_x is the fluid velocity in the x-direction; b denotes the channel width; ∇P_x denotes the pressure gradient in the x-direction; and a is the dynamic viscosity of the fluid.

Case 2 Validation of Pure Diffusion Process Within the Coal Matrix: For this validation, a 2D diffusion model with dimensions $L_x \times L_y = 100 \times 100$ is constructed using the D2Q5 lattice scheme. A spherical particle with radius $r_s = 30$ is placed at the center of the domain. The gas concentration in the region external to the spherical particle is maintained constant at $C_0 = 1$. Driven by the concentration gradient, gas diffuses radially inward from the spherical particle surface toward its interior. The analytical expression for the temporal evolution of the diffused mass within the spherical particle over time t is given by Eq. (3.48) [190]:

$$\frac{q}{q_0} = 1 - \sum_{n=1}^{\infty} \frac{4}{r_s^2 \alpha_n^2} \exp\left(-D_s \alpha_n^2 t\right) \tag{3.48}$$

where q denotes the mean adsorbed gas concentration within the particle; r_s denotes the particle radius; D_s is the gas diffusion coefficient inside the particle; and α_n are the positive roots of the zero-order Bessel function of the first kind.

The simulation parameters for the two validation models are detailed in Tables 3.2 and 3.3, with corresponding results presented in Figs. 3.4 and 3.5. The results demonstrate excellent agreement between the numerical simulations and the analytical solutions given by Eqs. (3.47) and (3.48).

Table 3.2 Simulation parameters of the validated model simulation of gas flow between matrices

L_x	L_y	u_{max}	∇P	μ
10	100	0.01	1.4171×10^{-6}	1/6

Table 3.3 Simulation parameters of the gas diffusion validation model within the matrix

L_x	L_y	r_s	C_0	D_s
100	100	30	1	0.03

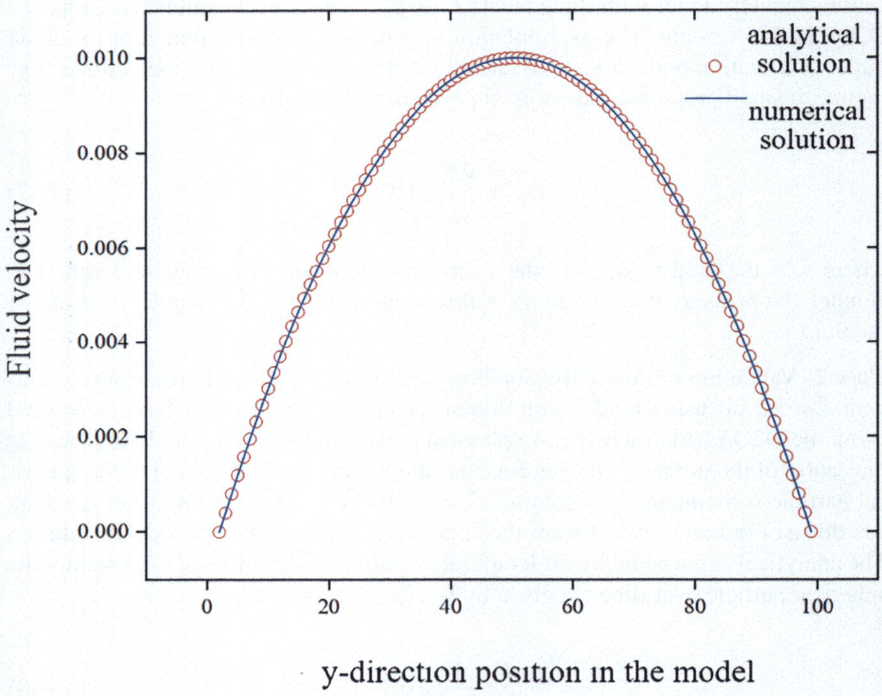

Fig. 3.4 Flow velocity at different locations in the y-direction space of the model

3.3.3 *Model Parameter Configuration*

The random circles packing (RCP) algorithm [191] is employed to reconstruct the coal matrix, thereby generating a stochastic porous medium that simulates the complex pore structure of coal. This reconstructed geometry is illustrated in Fig. 3.6.

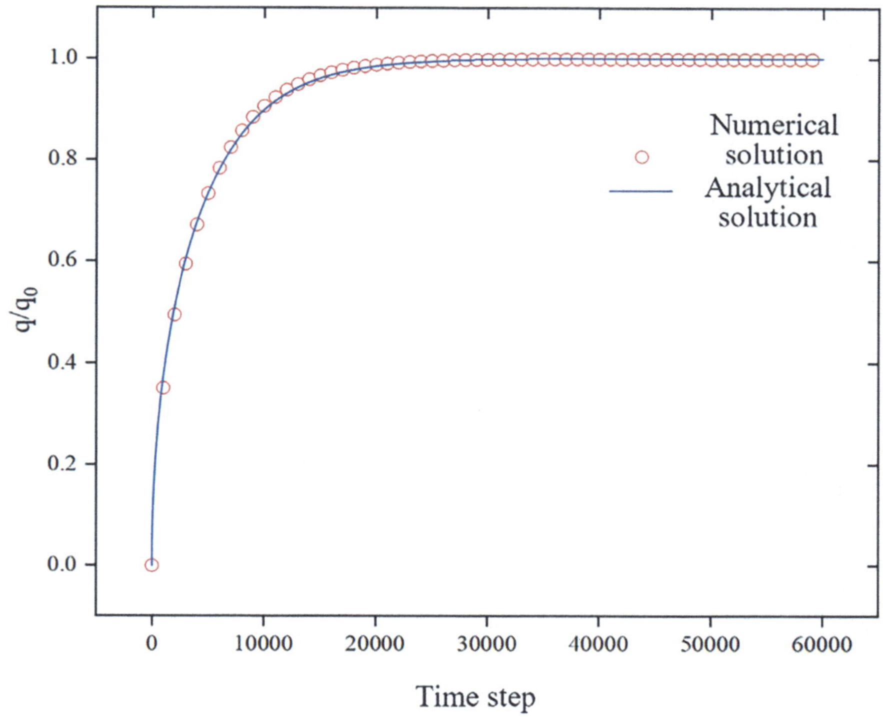

Fig. 3.5 Gas adsorption within spherical particles

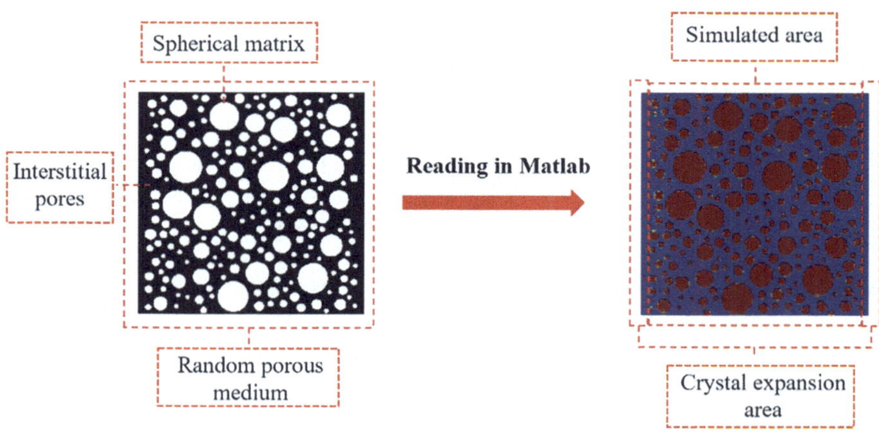

Fig. 3.6 Schematic of NSE-ADE coupled LB model construction

The 2D image of the stochastic porous medium in Fig. 3.6 was binarized using MATLAB software. In the processed image, black regions represent spherical coal matrix particles (assigned a value of 1). White regions represent pores (assigned a value of 0). To balance computational accuracy with simulation speed while ensuring sufficient resolution at the gas–solid interface, the simulation domain dimensions were set to $L_x \times L_y = 400 \times 400$ lattice units (i.e., a lattice resolution of 400 × 400). The physical pixel size was defined as 5 nm (i.e., $\Delta x = \Delta y = 5$ nm). Additionally, 10 extra lattice columns were appended to both the inlet and outlet boundaries to facilitate the implementation of the specified boundary schemes. The specific simulation parameters configured within the computational domain are detailed in Table 3.4.

Table 3.4 Analog parameter settings

Parameter	Symbol	Actual physical value	Lattice value
Simulation domain size	Lx	2×10^{-6} m	400
	Ly		400
Pixel size	Δx	5×10^{-9} m	1
	Δy		1
Time step	Δt	1.19×10^{-11} s	1
Fluid kinematic viscosity	μ	5.54×10^{-7} m²/s	0.264
Saturated gas concentration	C_{CH4}	0.75 kmol/m³	1
	C_{CO2}	2.31 kmol/m³	3.08
	C_{N2}	1.73 kmol/m³	2.31
Free-space gas diffusion coefficient	D_{CH4}	3.26×10^{-7} m²/s	0.155
	D_{CO2}	3.26×10^{-7} m²/s	0.155
	D_{N2}	3.26×10^{-7} m²/s	0.155
Gas diffusion coefficient within coal matrix	D_{sCH4}	1.08×10^{-9} m²/s	0.0051
	D_{sCO2}	1.52×10^{-9} m²/s	0.0072
	D_{sN2}	5.2×10^{-10} m²/s	0.0025
Adsorption rate constant	K_{aCH4}	2.21×10^{9} m³/kmol/s	0.0197
	K_{aCO2}	2.21×10^{9} m³/kmol/s	0.0197
	K_{aN2}	2.21×10^{9} m³/kmol/s	0.0197
Desorption rate constant	K_{dCH4}	3.78×10^{7} s^{-1}	0.00045
	K_{dCO2}	1.66×10^{7} s^{-1}	0.00020
	K_{dN2}	7.73×10^{7} s^{-1}	0.00092
Monolayer adsorption capacity	V_{mCH4}	3.08 kmol/m³	4.1
	V_{mCO2}	6 kmol/m³	8
	V_{mN2}	2.57 kmol/m³	3.43
	V^*_{mCH4}	0.45 kmol/m³	0.6

Note: Gas-related data in this table are sourced from Refs. [186–192] and the NIST Chemistry WebBook [193]

3.3.4 Boundary Conditions

The D2Q9-LB lattice model and D2Q5-LB lattice model are employed to solve the Navier-Stokes equations for the fluid velocity field and the advection-diffusion equations for the multicomponent concentration field, respectively. Distinct boundary conditions are applied to each lattice model as schematically shown in Fig. 3.7. For the Navier-Stokes lattice model, a constant injection velocity u_0 is imposed at the inlet (left boundary) for the injected gas. The bounce-back scheme is applied at the top/bottom boundaries and coal matrix interfaces. The outlet (right boundary) adopts the extrapolation scheme with a constant pressure p_0. For the advection-diffusion lattice model, the injected gas (CO_2 or N_2) concentration is set to a constant value C_0 at the inlet, while the CH_4 concentration is fixed at 0. The bounce-back boundary is implemented for all gases at the top/bottom walls, indicating no adsorption occurs. Langmuir adsorption/desorption dynamics govern gas interactions at coal matrix surfaces for both injected gases and CH_4. The outlet uses a zero-concentration-gradient boundary condition. A summary of all boundary conditions utilized in the simulations is provided in Table 3.5.

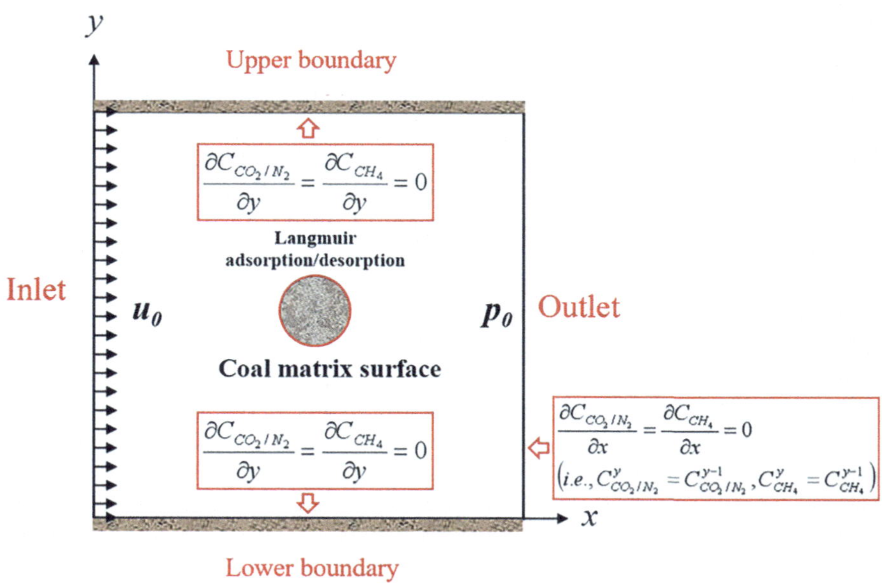

Fig. 3.7 Schematic of CH_4 boundary conditions in gas injection displaced coal

Table 3.5 NSE-ADE coupled LB model boundary conditions

Location	Navier-Stokes lattice model	Advection-diffusion lattice model	
		CO_2/N_2	CH_4
Left	Constant velocity, u_0	Constant concentration, C_0	Constant concentration, 0
Right	Constant pressure, p_0	$C^{y}_{CO_2/N_2} = C^{y-1}_{CO_2/N_2}$	$C^{y}_{CH_4} = C^{y-1}_{CH_4}$
Top	Bounce-back	Bounce-back	Bounce-back
Bottom	Bounce-back	Bounce-back	Bounce-back
Coal matrix interface	Bounce-back	Langmuir adsorption/desorption dynamics	Langmuir adsorption/desorption dynamics

3.4 Dynamic Transport Behavior of CH_4 During Gas Injection Displacement in Coal

3.4.1 Impact of Injection Gas Type on CH_4 Transport Dynamics

3.4.1.1 Evolution of Total Adsorbed Gas and Free CH_4 Content

Figures 3.8 and 3.9 present simulation snapshots of adsorbed amounts for hetero-component gases and CH_4 within the coal spherical particle porous medium at 0.1, 0.5, 1.25, and 2.0 μs. During CO_2 injection displacement, CO_2 rapidly adsorbs onto coal matrix surfaces even at low initial concentrations. As displacement progresses, interparticle CO_2 concentration increases, with CO_2 gradually diffusing into the coal matrix interior. Concurrently, free CH_4 migrates toward the outlet driven by pressure gradients, reducing interparticle CH_4 concentration. This decline triggers desorption: adsorbed CH_4 on matrix surfaces and within internal pores diffuses into interparticle spaces under concentration gradients.

In contrast, N_2 injection exhibits significantly weaker initial adsorption. N_2 slowly adsorbs onto coal surfaces during continuous injection and gradually diffuses inward. Due to N_2's lower adsorption affinity than CH_4, competitive adsorption is absent. The injected N_2 elevates total system pressure while reducing CH_4 partial pressure, promoting desorption of adsorbed-phase CH_4 from surfaces and internal pores, which then diffuses outward along concentration gradients. This mechanistic difference fundamentally arises from CO_2's strong competitive adsorption versus N_2's partial-pressure-driven displacement.

Figure 3.10 illustrates the temporal evolution of adsorbed amounts for hetero-component gases and CH_4 within the coal matrix during displacement. The adsorbed mass of injected gas exhibits an initial rapid increase followed by a gradual rise before stabilizing. Due to coal's stronger affinity for CO_2 adsorption compared to N_2, CO_2 demonstrates faster accumulation, higher final adsorbed mass, and a longer stabilization period. Conversely, CH_4 adsorption decreases rapidly initially, then declines gradually before stabilizing under both injection scenarios. CO_2 injection

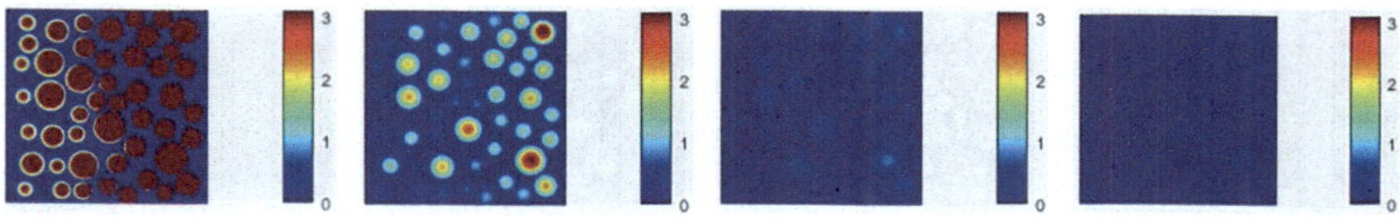

(a) Simulation snapshots of CH_4 adsorption, from left to right: 0.1 μs、0.5 μs、1.25 μs、2.0 μs

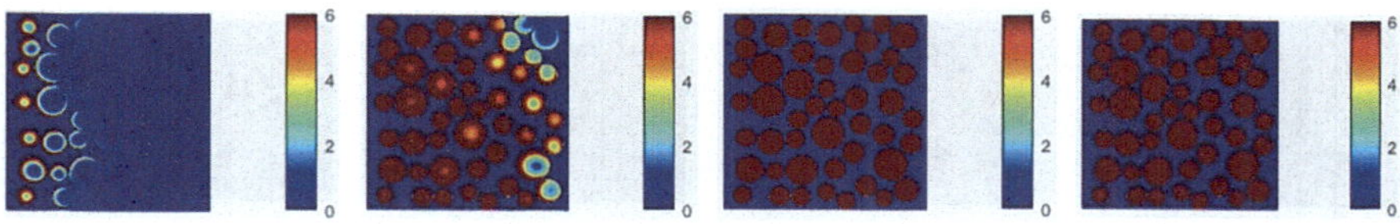

(b) Simulation snapshots of CO_2 adsorption, from left to right: 0.1 μs、0.5 μs、1.25 μs、2.0 μs

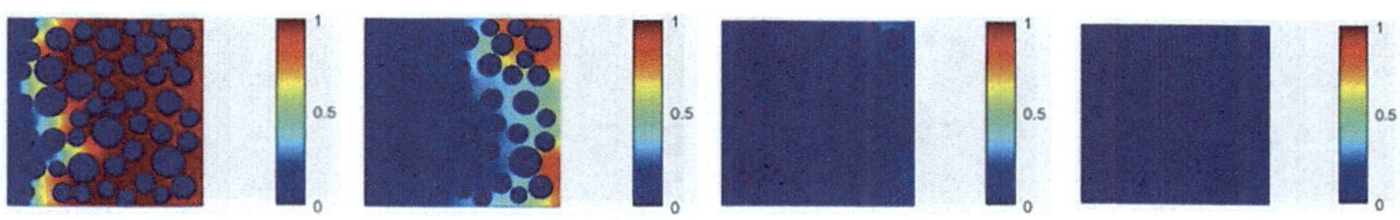

(c) Simulation snapshots of free-state CH_4, from left to right: 0.1 μs、0.5 μs、1.25 μs、2.0 μs

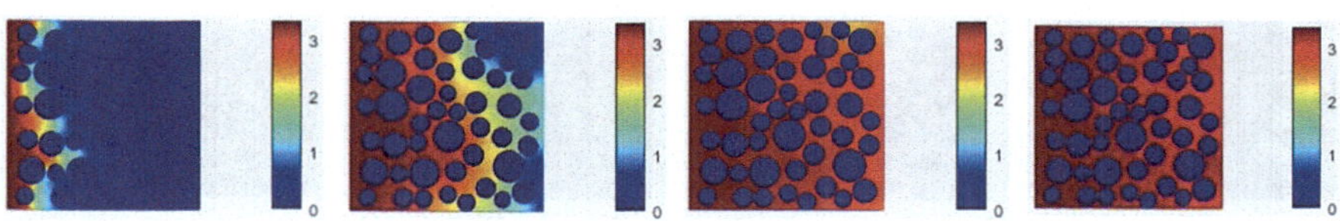

(d) Simulation snapshots of free-state CH_4, from left to right: 0.1 μs、0.5 μs、1.25 μs、2.0 μs

Fig. 3.8 Snapshots of simulation of gas adsorption and free volume during CO_2 injection and replacement process. (**a**) Simulation snapshots of CH_4 adsorption, from left to right: 0.1, 0.5, 1.25, and 2.0 μs. (**b**) Simulation snapshots of CO_2 adsorption, from left to right: 0.1, 0.5, 1.25, and 2.0 μs. (**c**) Simulation snapshots of free-state CH_4, from left to right: 0.1, 0.5, 1.25, and 2.0 μs. (**d**) Simulation snapshots of free-state CH_4, from left to right: 0.1, 0.5, 1.25, and 2.0 μs

induces significantly faster CH_4 desorption and earlier stabilization than N_2 injection. This divergence stems from the adsorption strength hierarchy: $CO_2 > CH_4 > N_2$, where CO_2's competitive adsorption actively displaces adsorbed CH_4.

Figure 3.11 depicts the evolution of free CH_4 content in interparticle spaces during gas injection displacement. Overall, the trends under CO_2 and N_2 injection exhibit fundamental similarities, though nuanced differences manifest in the phase-specific desorption kinetics. During the initial displacement stage, N_2 injection achieves a faster reduction in free CH_4 compared to CO_2 injection. This occurs due to N_2's weak adsorption affinity, which enhances its "sweeping effect"—efficiently displacing free-phase CH_4 from the computational domain. As displacement progresses, while the free CH_4 level under N_2 injection remains lower than with CO_2, the decline rate accelerates significantly under CO_2 injection. This shift arises

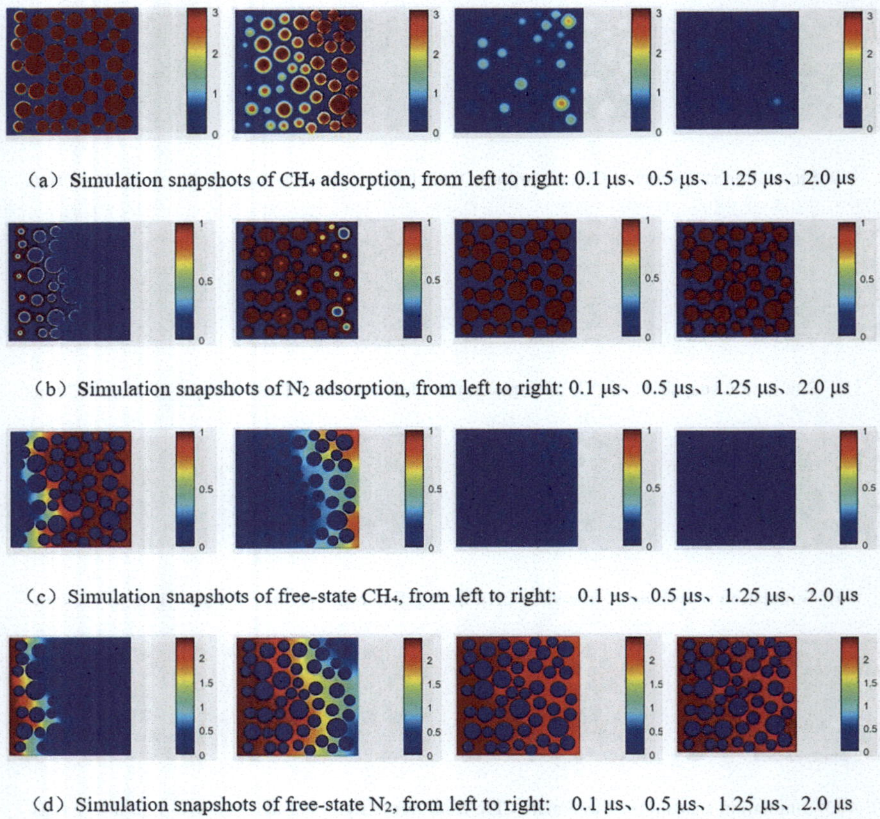

Fig. 3.9 Snapshot of the simulation of gas adsorption and free volume during N_2 injection and replacement process. (**a**) Simulation snapshots of CH_4 adsorption, from left to right: 0.1, 0.5, 1.25, and 2.0 μs. (**b**) Simulation snapshots of N_2 adsorption, from left to right: 0.1, 0.5, 1.25, and 2.0 μs. (**c**) Simulation snapshots of free-state CH_4, from left to right: 0.1, 0.5, 1.25, and 2.0 μs. (**d**) Simulation snapshots of free-state N_2, from left to right: 0.1, 0.5, 1.25, and 2.0 μs

because adsorbed CH_4 begins desorbing from matrix surfaces and internal pores; CO_2's stronger competitive adsorption liberates more adsorbed CH_4 per unit time, which is subsequently swept out. Given the fixed initial adsorbed CH_4 reservoir, this accelerated desorption causes a steeper free CH_4 decrease under CO_2 injection. In the advanced stage, free CH_4 becomes primarily replenished by CH_4 diffusing from the coal matrix interior. Since CO_2 injection has depleted more matrix-reserved CH_4 than N_2 injection, less CH_4 migrates into interparticle spaces per unit time, consequently slowing the free CH_4 decline rate under CO_2. Ultimately, the free CH_4 disparity between the two injection scenarios narrows and converges toward equilibrium.

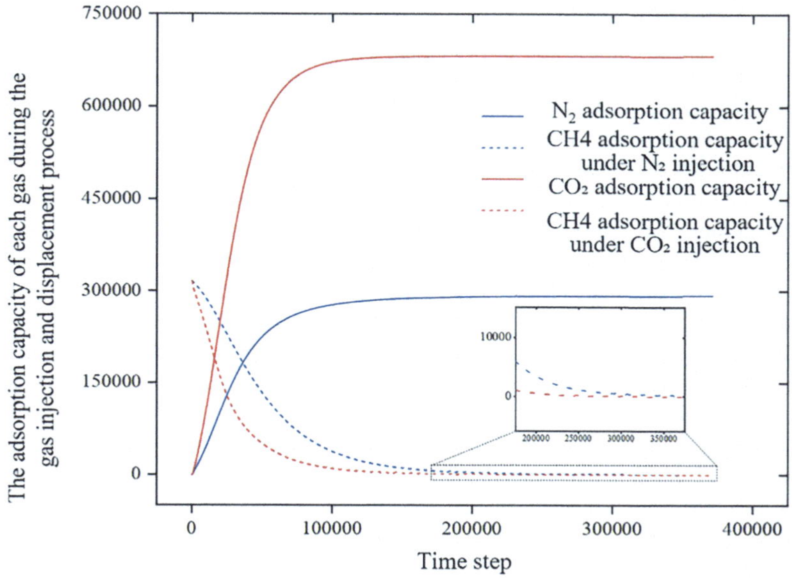

Fig. 3.10 Variation of adsorption of heterogeneous and CH_4 gases in the coal matrix with time of expulsion

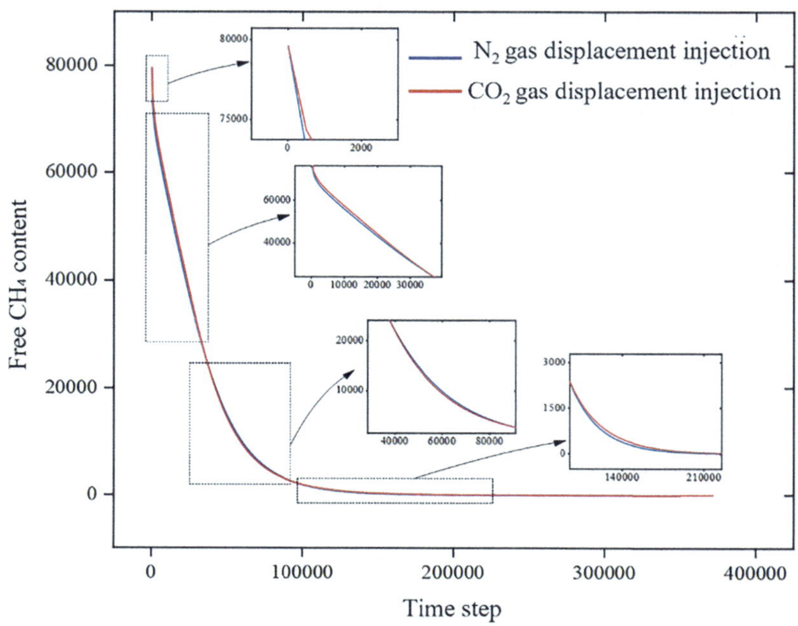

Fig. 3.11 Variation of free CH_4 gas between coal matrices as a function of time of expulsion

3.4.1.2 CH_4 Outlet Concentration Evolution

Figure 3.12 depicts the temporal variation of CH_4 concentration at the outlet during gas injection displacement. During the initial displacement phase (0–0.02 μs), coal exhibits weaker adsorption affinity for N_2 than CO_2, resulting in a higher free-phase N_2 content within the computational domain under N_2 injection. Consequently, compared to CO_2 injection, N_2 achieves a more efficient "sweeping effect," displacing a larger volume of pre-existing free-phase CH_4. This mechanism drives a faster decline in CH_4 outlet concentration under N_2 injection. Notably, >90% of the outlet CH_4 flux during this stage originates from pre-existing free-phase CH_4 displacement.

During the early displacement stage (0.02–0.28 μs), adsorbed CH_4 desorbs and replenishes interparticle spaces, causing a slight rise in CH_4 outlet concentration. Due to CO_2's superior ability to liberate adsorbed CH_4 per unit time compared to N_2, CO_2 injection drives a significantly steeper increase in outlet CH_4 concentration.

In the mid-term stage (0.28–1.0 μs), virtually no pre-existing free CH_4 remains. Adsorbed CH_4 from matrix surfaces and near-inlet regions desorbs into macropores. CO_2's enhanced desorption capacity elevates outlet CH_4 concentration above N_2-injection levels, primarily sustained by surface-adsorbed CH_4. However, as this surface reservoir is finite, CO_2 injection exhibits a faster decline rate in outlet concentration than N_2.

During the late displacement stage (>1.0 μs), deep-matrix CH_4 becomes the dominant source for outlet flux, where surface-adsorbed CH_4 is defined as readily

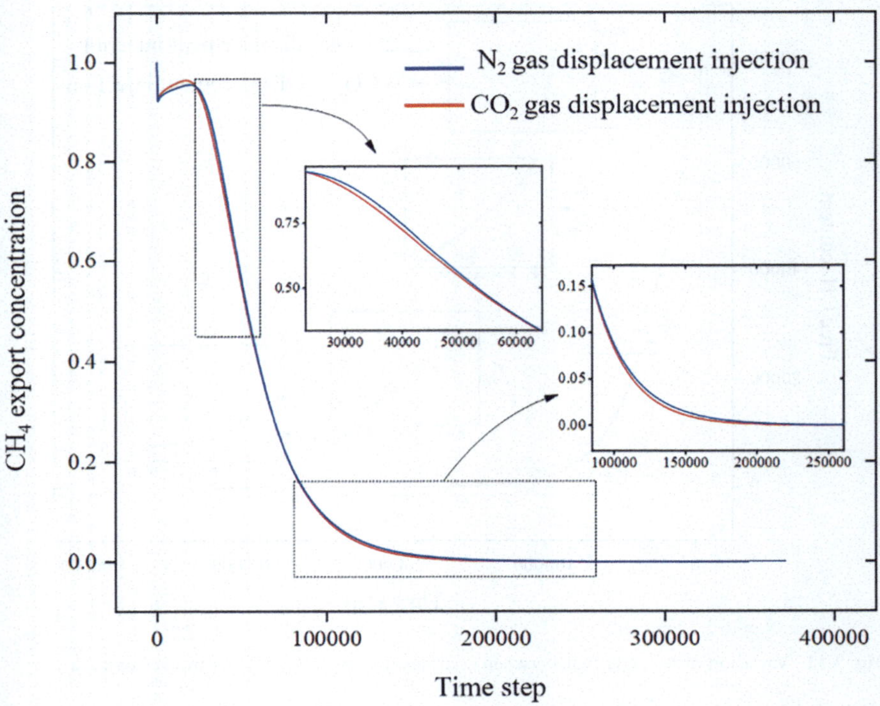

Fig. 3.12 Variation of CH_4 gas concentration at the outlet during gas injection drive with drive time

accessible and matrix-interior CH_4 as less accessible. CO_2 outperforms N_2 in desorbing this deeply trapped CH_4, enabling faster desorption of matrix-interior CH_4 per unit time under CO_2 injection. Consequently, the outlet CH_4 concentration declines more steeply compared to N_2 injection, though the disparity between the two methods progressively narrows until equilibration due to reservoir depletion constraints.

3.4.1.3 Evolution of CH_4 Production Volume

Figure 3.13 illustrates the evolution of CH_4 emissions during gas injection displacement. The cumulative CH_4 emissions exhibit three distinct phases: an initial rapid linear increase, followed by a decelerating growth phase, and eventual stabilization. In the rapid-increase phase, emissions rise at a constant rate, predominantly contributed by displacement of free-phase CH_4 in interparticle spaces. Transitioning to the deceleration phase, the emission growth rate gradually declines as the primary source shifts to desorption of readily accessible CH_4 adsorbed on matrix surfaces and near-injection boundaries. During the final stabilization phase, emissions plateau as they become sustained solely by desorption of less accessible CH_4 from deep-matrix regions. Comparatively, CO_2 injection achieves both a higher emission growth rate and greater cumulative emissions than N_2 injection, attributable to CO_2's superior competitive adsorption capacity that enhances CH_4 liberation efficiency throughout all displacement stages.

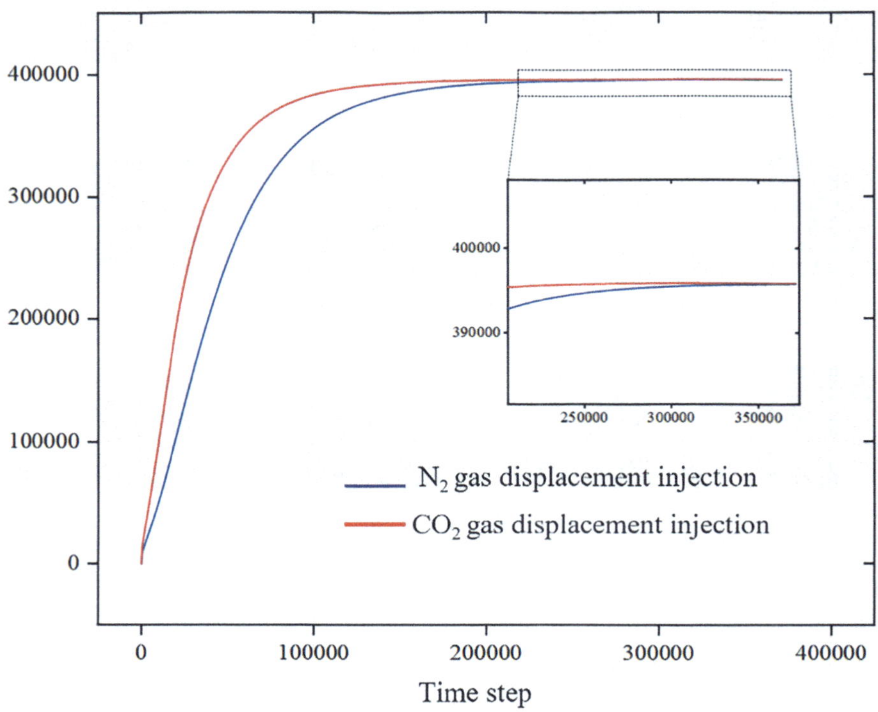

Fig. 3.13 Variation of CH_4 gas concentration at the outlet during gas injection drive with drive time

3.4.2 Impact of Injection Flow Rate on CH_4 Dynamic Transport Behavior

Using CO_2 injection for coalbed CH_4 displacement as a case study, this section investigates how injection flow rates influence CH_4 transport dynamics.

3.4.2.1 Evolution of Total Adsorbed Gas and Free CH_4 Content

Figures 3.14, 3.15 and 3.16 present simulation snapshots of heterocomponent gas adsorption and free-phase distributions under lattice-scaled injection flow rates of 0.05, 0.10, and 0.20. It is evident that during the initial displacement stage, higher

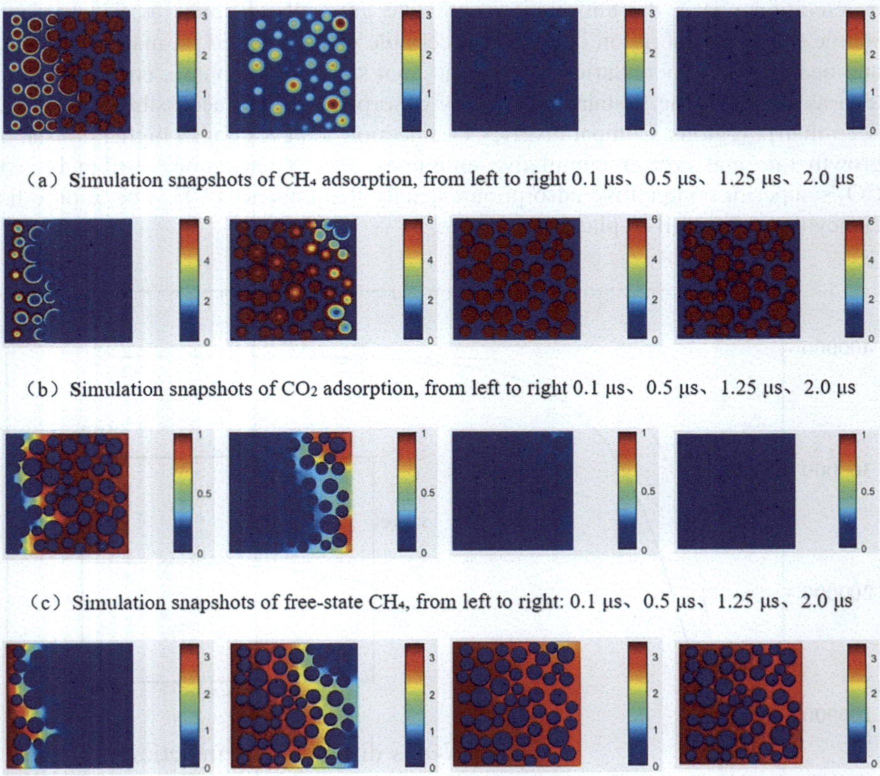

(a) Simulation snapshots of CH_4 adsorption, from left to right 0.1 μs、0.5 μs、1.25 μs、2.0 μs

(b) Simulation snapshots of CO_2 adsorption, from left to right 0.1 μs、0.5 μs、1.25 μs、2.0 μs

(c) Simulation snapshots of free-state CH_4, from left to right: 0.1 μs、0.5 μs、1.25 μs、2.0 μs

(d) Simulation snapshots of free-state CO_2, from left to right: 0.1 μs、0.5 μs、1.25 μs、2.0 μs

Fig. 3.14 Snapshots of gas adsorption and free volume simulations at an injection flow rate of 0.05. (**a**) Simulation snapshots of CH_4 adsorption, from left to right: 0.1, 0.5, 1.25, and 2.0 μs. (**b**) Simulation snapshots of CO_2 adsorption, from left to right: 0.1, 0.5, 1.25, and 2.0 μs. (**c**) Simulation snapshots of free-state CH_4, from left to right: 0.1, 0.5, 1.25, and 2.0 μs. (**d**) Simulation snapshots of free-state CO_2, from left to right: 0.1, 0.5, 1.25, and 2.0 μs

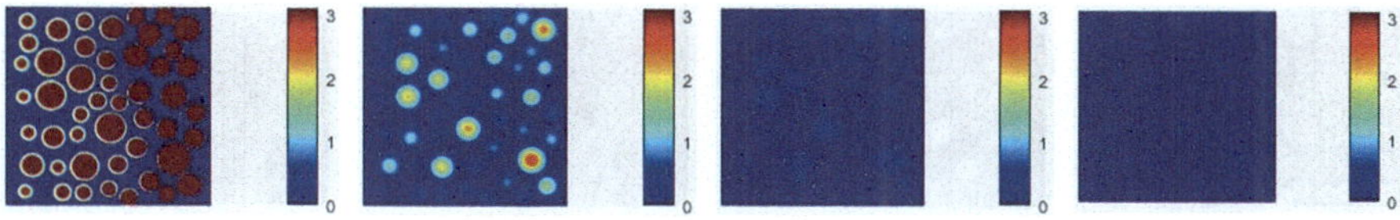

(a) Simulation snapshots of CH_4 adsorption, from left to right: 0.1 μs、0.5 μs、1.25 μs、2.0 μs

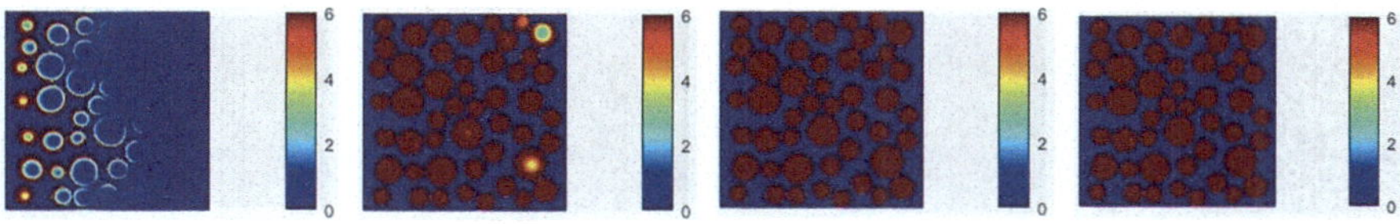

(b) Simulation snapshots of CO_2 adsorption, from left to right: 0.1 μs、0.5 μs、1.25 μs、2.0 μs

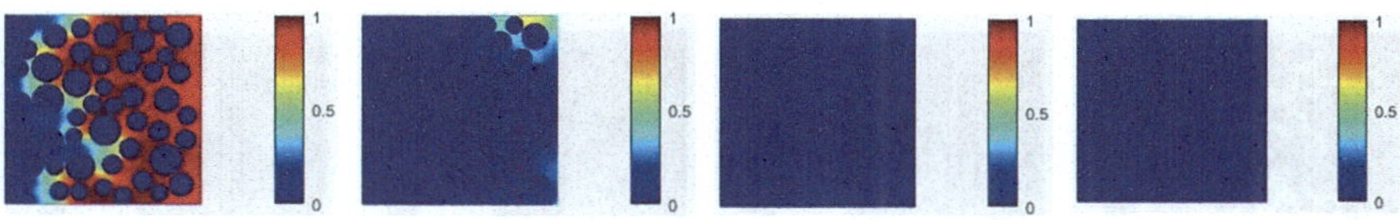

(c) Simulation snapshots of free-state CH_4, from left to right: 0.1 μs、0.5 μs、1.25 μs、2.0 μs

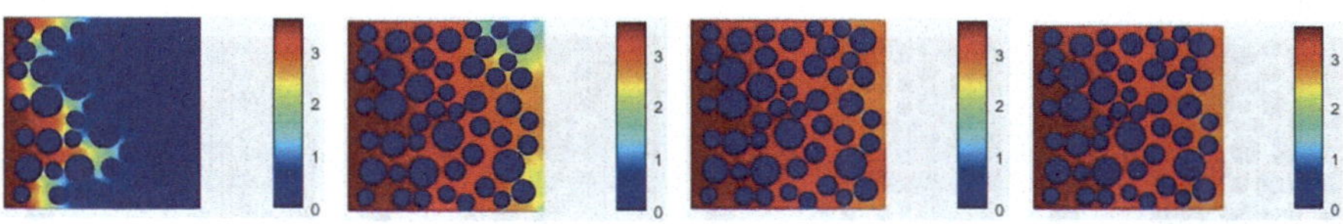

(d) Simulation snapshots of free-state CO_2, from left to right: 0.1 μs、0.5 μs、1.25 μs、2.0 μs

Fig. 3.15 Snapshots of gas adsorption and free volume simulations at an injection flow rate of 0.10. (**a**) Simulation snapshots of CH_4 adsorption, from left to right: 0.1, 0.5, 1.25, and 2.0 μs. (**b**) Simulation snapshots of CO_2 adsorption, from left to right: 0.1, 0.5, 1.25, and 2.0 μs. (**c**) Simulation snapshots of free-state CH_4, from left to right: 0.1, 0.5, 1.25, and 2.0 μs. (**d**) Simulation snapshots of free-state CO_2, from left to right: 0.1, 0.5, 1.25, and 2.0 μs

injection flow rates correlate with expanded spatial extents of CH_4 desorption and heterocomponent gas adsorption across the computational domain at identical time points. Concurrently, the propagation front of free-phase heterocomponent gas advances farther, while the spatial distribution of free-phase CH_4 contracts more rapidly, confirming that elevated flow rates accelerate the displacement of free CH_4. As displacement progresses, higher flow rates yield lighter particle coloration (indicating reduced CH_4 adsorption) and darker particle coloration (signifying enhanced heterocomponent gas adsorption) at equivalent timestamps, demonstrating that increased injection rates intensify heterocomponent gas adsorption and accelerate CH_4 desorption through competitive mechanisms.

Figure 3.17 illustrates the temporal evolution of adsorbed amounts for heterocomponent gases and CH_4 under lattice-scaled injection flow rates of 0.05, 0.10, and 0.20. Universally, heterocomponent gas adsorption exhibits an initial rapid increase

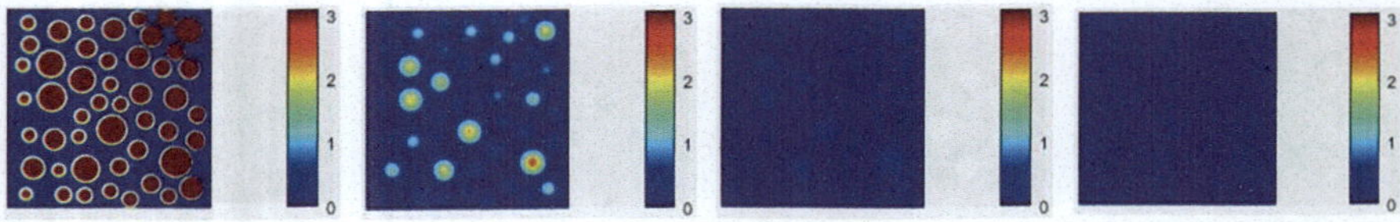

(a) Simulation snapshots of CH_4 adsorption, from left to right: 0.1 μs、0.5 μs、1.25 μs、2.0 μs

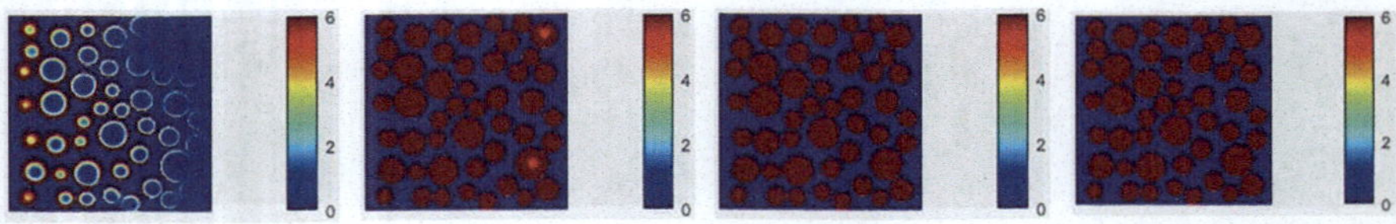

(b) Simulation snapshots of CO_2 adsorption, from left to right: 0.1 μs、0.5 μs、1.25 μs、2.0 μs

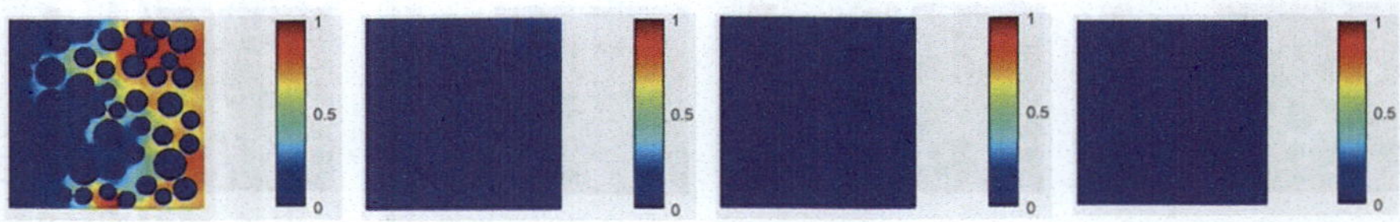

(c) Simulation snapshots of free-state CH_4, from left to right: 0.1 μs、0.5 μs、1.25 μs、2.0 μs

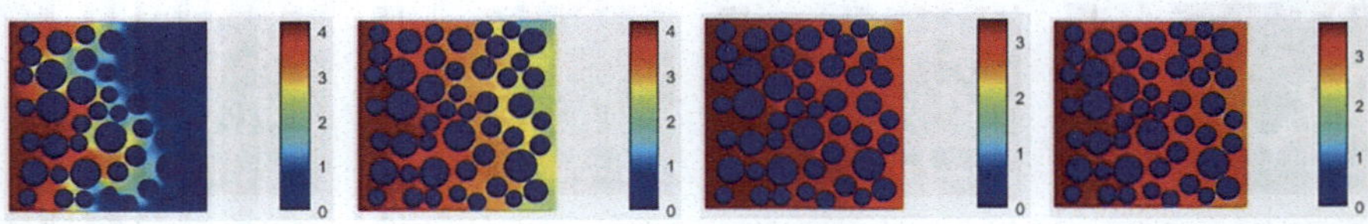

(d) Simulation snapshots of free-state CO_2, from left to right: 0.1 μs、0.5 μs、1.25 μs、2.0 μs

Fig. 3.16 Snapshots of gas adsorption and free volume simulations at an injection flow rate of 0.20. (**a**) Simulation snapshots of CH_4 adsorption, from left to right: 0.1, 0.5, 1.25, and 2.0 μs. (**b**) Simulation snapshots of CO_2 adsorption, from left to right: 0.1, 0.5, 1.25, and 2.0 μs. (**c**) Simulation snapshots of free-state CH_4, from left to right: 0.1, 0.5, 1.25, and 2.0 μs. (**d**) Simulation snapshots of free-state CO_2, from left to right: 0.1, 0.5, 1.25, and 2.0 μs

followed by gradual growth before stabilization, while CH_4 adsorption decreases rapidly initially, then declines gradually before stabilizing. Comparatively, higher flow rates accelerate heterocomponent adsorption growth and intensify CH_4 adsorption decline. This stems from two mechanisms: (1) Elevated flow rates deliver greater heterocomponent gas mass per unit time, amplifying competitive adsorption and partial pressure effects on adsorbed CH_4 and (2) increased injected gas flux enlarges concentration gradients between adsorbed-phase CH_4 and free-phase reservoirs. Consequently, higher flow rates boost heterocomponent adsorption per unit time, accelerate CH_4 desorption rates, and shorten the time to adsorption equilibrium for all gases.

Figure 3.18 depicts the evolution of free CH_4 content across flow rates. Free CH_4 universally declines rapidly initially, then gradually before stabilizing. Higher injection rates accelerate this decline and reduce the time to stabilization, as increased

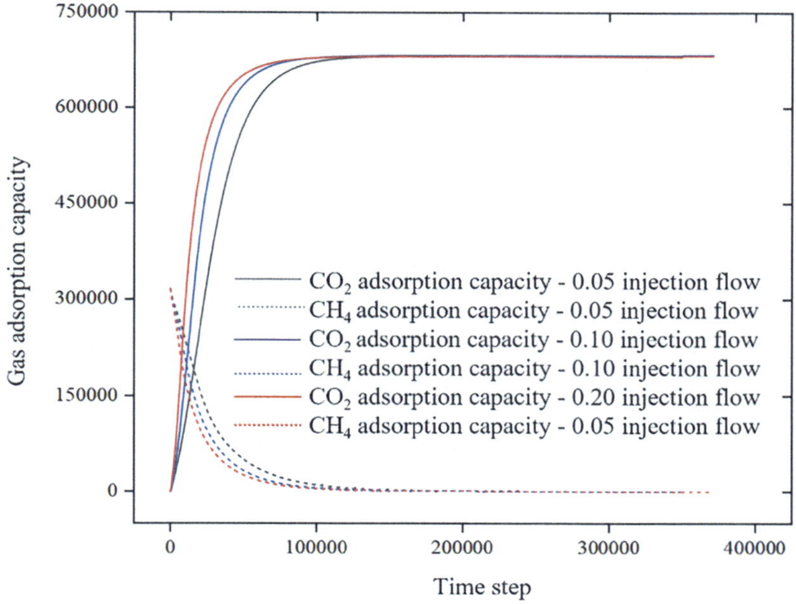

Fig. 3.17 Variation of adsorption of iso-elemental gas with CH_4 gas as a function of drive time for each injection flow rate condition

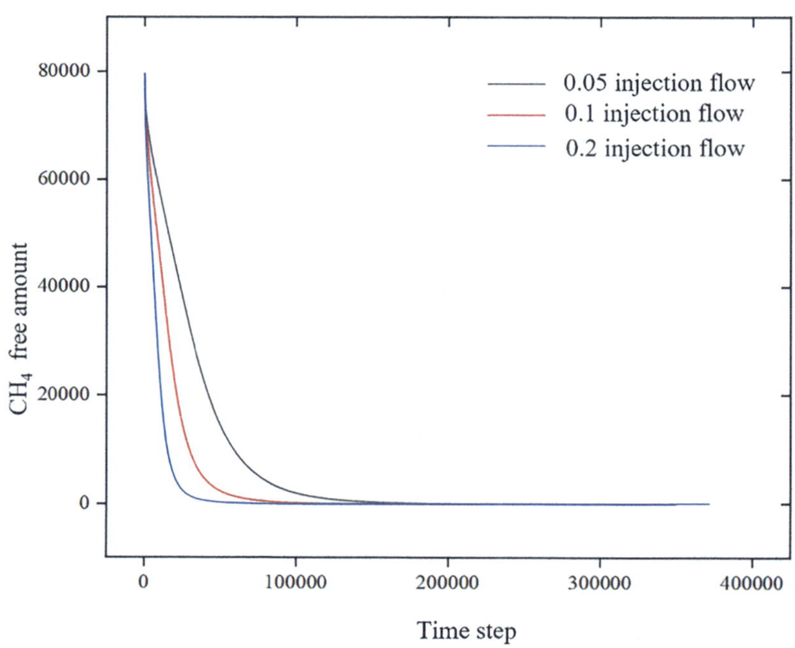

Fig. 3.18 Variation of CH_4 gas free with time of exhumation for each injection flow rate condition

advective flux enhances displacement efficiency of free-phase CH_4 while intensified desorption (Fig. 3.17) diminishes replenishment from adsorbed reservoirs.

3.4.2.2 Evolution of CH_4 Outlet Concentration

Figure 3.19 depicts the temporal variation of CH_4 concentration at the outlet under varying injection flow rates.

Higher injection flow rates deliver greater mass flux of heterocomponent gas per unit time, resulting in steeper initial declines in CH_4 outlet concentration during the early displacement phase. However, as established previously, elevated flow rates simultaneously accelerate CH_4 desorption. This desorbed CH_4 replenishes the free-phase reservoir, inducing a transient rise in outlet concentration—an effect amplified at higher flow rates ($0.20 > 0.10 > 0.05$). As displacement progresses, increased flow rates drive faster declines in CH_4 outlet concentration and shorten the time to stabilization due to intensified reservoir depletion.

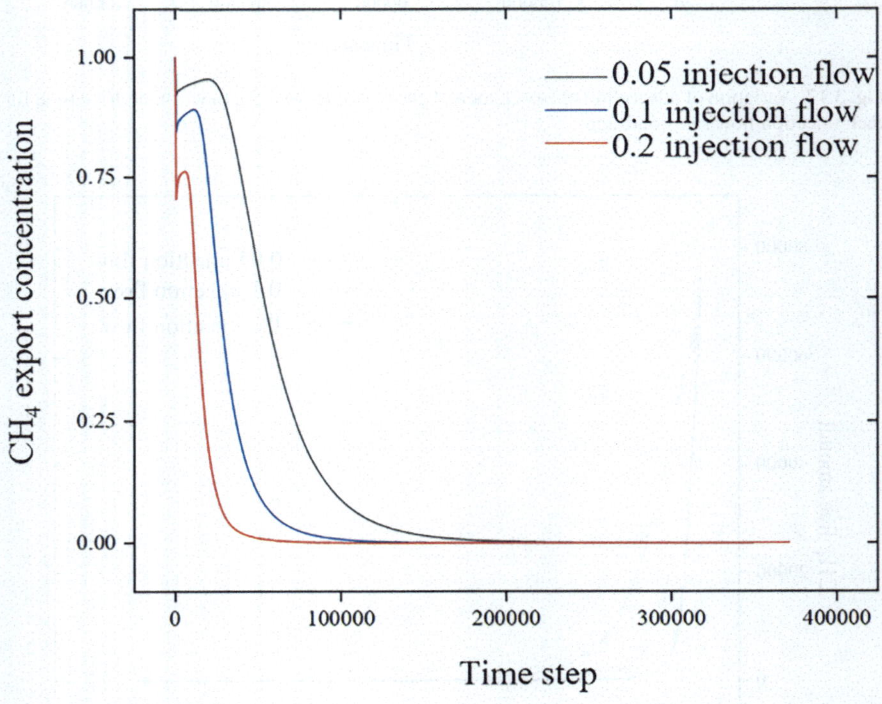

Fig. 3.19 Variation of CH_4 gas concentration at the outlet with drive time for each injection flow condition

3.4.2.3 Evolution of CH_4 Production Volume

Figure 3.20 illustrates the temporal evolution of cumulative CH_4 production volume under varying injection flow rates.

Higher injection flow rates accelerate the growth rate of cumulative CH_4 production volume and shorten the time to stabilization. This acceleration stems from intensified displacement mechanisms: (1) enhanced entrainment effect of hetero-component gas on free-phase CH_4 and (2) amplified competitive adsorption and partial pressure effects on adsorbed-phase CH_4. Consequently, greater CH_4 volumes are displaced per unit time. Ultimately, total cumulative production converges across flow rates, demonstrating that injection flow rate primarily governs the production velocity rather than the recoverable reserve capacity, as constrained by finite in situ CH_4 resources.

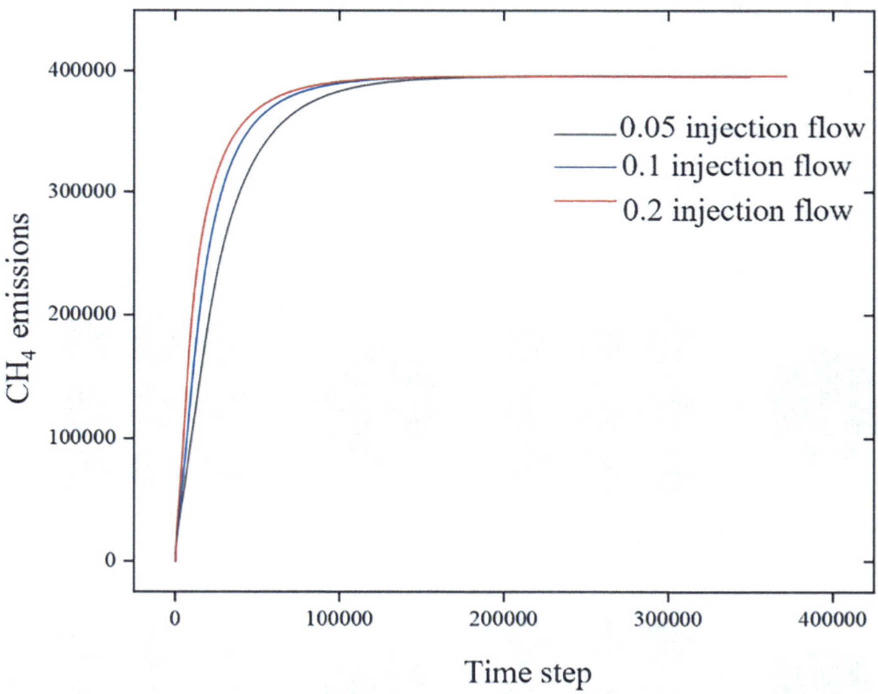

Fig. 3.20 Variation of CH_4 gas emissions with replacement time for each injection flow condition

3.5 Migration-Production Mechanism of CH_4 in Coal during Gas Injection Displacement

The dynamic transport of CH_4 during gas injection displacement progresses through three distinct stages (I, II, and III), as illustrated in Fig. 3.21. Initially, CH_4 exists in adsorption equilibrium across three domains: free-phase CH_4 in interparticle spaces, adsorbed-phase CH_4 on matrix surfaces, and adsorbed-phase CH_4 within the matrix interior.

Stage I commences with gas injection, where the injected heterocomponent gas elevates the pressure differential between the coal matrix interparticle spaces and the outlet boundary. This pressure gradient drives rapid outflow of free-phase CH_4, causing a sharp decline in interparticle free CH_4 content—a rate modulated by injection flow rate and gas type (Figs. 3.11 and 3.18). Concurrently, a transient rise in outlet CH_4 concentration occurs, its magnitude governed by injection parameters (Figs. 3.12 and 3.19). While minor desorption of adsorbed-phase CH_4 occurs near the injection boundary, >85% of outlet CH_4 flux originates from the pre-existing free-phase reservoir.

Stage II initiates when free-phase CH_4 is depleted. Heterocomponent gases now trigger bulk desorption of surface-adsorbed CH_4 through competitive adsorption and partial pressure reduction, while simultaneously adsorbing onto matrix surfaces. Desorption and adsorption kinetics are controlled by gas type and flow rate (Figs. 3.10 and 3.17). Adsorbed CH_4 from subsurface layers near

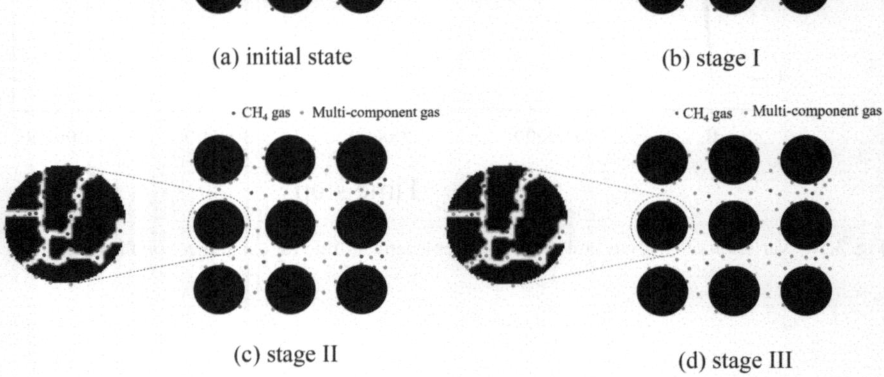

Fig. 3.21 Schematic diagram of CH_4 transport in the coal during gas injection and displacement. (**a**) Initial state. (**b**) Stage I. (**c**) Stage II. (**d**) Stage III

the injection boundary desorbs and diffuses into macropores, yet outlet CH_4 primarily derives from readily accessible surface-adsorbed CH_4. This phase exhibits accelerated declines in adsorbed CH_4, free interparticle CH_4, and outlet concentration.

Stage III emerges as heterocomponent gases penetrate the matrix interior, mobilizing deeply trapped adsorbed-phase CH_4 (classified as less accessible) into interparticle spaces. Due to lengthened diffusion pathways through the coal matrix structure, the decline rates of adsorbed CH_4, free interparticle CH_4, and outlet concentration progressively decelerate before stabilizing (Figs. 3.10, 3.11 and 3.12). During this terminal phase, outlet CH_4 flux primarily comprises CH_4 liberated from the matrix interior, completing the three-stage migration-depletion cycle.

3.6 Chapter Summary

This chapter developed a coupled NSE-ADE lattice Boltzmann model incorporating directional flow of multicomponent gases and intra/inter-matrix mass transfer during gas injection displacement. Simulations investigated the dynamic transport behavior of CH_4, specifically examining the impacts of injection flow rate and gas type (CO_2 vs. N_2) on displacement dynamics. A comprehensive analysis of CH_4 migration-production mechanisms yielded the following conclusions:

1. The coupled NSE-ADE lattice Boltzmann model was developed to simulate directional flow of multicomponent gases and intra/inter-matrix mass transfer during gas injection displacement. Validation confirmed high-fidelity replication of interparticle gas flow and intraparticle pure diffusion, with simulation results showing excellent agreement ($R^2 > 0.98$) against analytical solutions.
2. The gas type significantly influences CH_4 transport dynamics during displacement: CO_2 exhibits faster adsorption kinetics and higher ultimate adsorption capacity than N_2 within the computational domain, requiring a longer stabilization period. Conversely, CH_4 adsorption declines more rapidly under CO_2 injection, reaching equilibrium faster than with N_2. For free-phase CH_4 content, N_2 injection achieves faster initial decline due to enhanced sweeping efficiency. During mid-displacement, free CH_4 levels remain lower under N_2 than CO_2 injection, but CO_2 drives a steeper decline rate. By late-stage, free CH_4 concentrations converge across both gases. Outlet CH_4 concentration shows three-phase variation: N_2 yields higher initial concentrations from free-phase displacement; CO_2 generates elevated mid/late-stage concentrations via competitive desorption; while late-stage concentrations under CO_2 remain higher but narrow toward convergence. Critically, CO_2 injection achieves both 28% faster cumulative CH_4 production growth and 22% greater ultimate yield compared to N_2, demonstrating superior recovery efficiency.

3. Injection flow rate critically modulates CH_4 transport dynamics: Elevated flow rates accelerate heterocomponent gas adsorption and amplify CH_4 desorption per unit time, thereby reducing the stabilization period for all gaseous species. Concurrently, free-phase CH_4 content declines more rapidly under higher flow rates, reaching stabilization faster. Outlet CH_4 concentration exhibits accelerated decay kinetics with increased flow rates, shortening the time to concentration stabilization. Similarly, cumulative CH_4 production growth rates intensify with higher injection fluxes, achieving production plateau earlier. Crucially, while flow rate governs process kinetics, total recoverable CH_4 remains invariant across tested flow conditions (0.05–0.20 lattice units), being constrained by in-place gas reserves rather than injection intensity.
4. The dynamic migration of CH_4 in coal under gas injection displacement progresses through three distinct stages: In stage I, injected heterocomponent gas elevates the pressure gradient between coal matrix interparticle spaces and the outlet boundary. Driven by this gradient, free-phase CH_4 rapidly flows toward the outlet, constituting >90% of effluent CH_4. In stage II, heterocomponent gases trigger bulk desorption of surface-adsorbed CH_4 via competitive adsorption and partial pressure reduction while adsorbing onto matrix surfaces. Effluent CH_4 primarily derives from readily accessible surface-adsorbed CH_4, accompanied by accelerated declines in adsorbed CH_4, free interparticle CH_4, and outlet concentration. In stage III, heterocomponent gases diffuse into the matrix interior, mobilizing less accessible adsorbed-phase CH_4 into interparticle spaces. Decline rates of adsorbed CH_4, free interparticle CH_4, and outlet concentration progressively decelerate due to diffusion-path constraints before stabilizing. Effluent CH_4 transitions to predominantly matrix-interior sourced CH_4, completing the depletion cycle.

Chapter 4
Experimental Investigation on Dynamic Transport Behavior of CH_4 in Coal Particles During Gas Injection Displacement

4.1 Introduction

Previous studies have clarified microscale CH_4 transport mechanisms during gas injection displacement and revealed molecular-level migration-production dynamics. However, at the macroscale, the dynamic transport behavior of CH_4 remains unexplored, and it is unknown whether macroscale manifestations align with microscale properties. Furthermore, existing theoretical models oversimplify diffusion processes by substituting the CH_4 diffusion coefficient with values measured under single-gas conditions—an approach lacking physical justification. Critical knowledge gaps persist regarding the applicability of unary gas diffusion models to CH_4's dynamic diffusion behavior during displacement, and the fundamental discrepancies between displacement-driven diffusion dynamics and single-gas diffusion characteristics demand rigorous investigation.

This chapter employs a self-developed experimental system for characterizing CH_4 dynamic transport during gas injection displacement to quantitatively investigate CH_4 migration patterns under varied conditions. Using macroscopic experimental data, we theoretically assess the applicability of unary gas diffusion models for representing CH_4's dynamic diffusion behavior during displacement, while examining how injection flow rates and gas types influence diffusion parameter variations across models. Furthermore, by comparing experimental results from single-gas CH_4 diffusion and displacement-driven CH_4 diffusion, we analyze fundamental divergences between these two diffusion regimes.

G. Wang, H. Gong, *Dynamic Migration Characteristics of CH4 During Gas Injection Displacement in Coal Seams*,
https://doi.org/10.1007/978-981-95-8879-4_4

4.2 Experimental System for Determining the Dynamic Transport Characteristics of CH_4 in Coal by Gas Injection Displacement

4.2.1 Construction of the Experimental System

A schematic diagram of the experimental platform is shown in Fig. 4.1. The apparatus consists of a gas injection and back-pressure system, a gas injection flow control and monitoring system, a coal sample loading system, a data acquisition system, a constant temperature chamber, and a vacuum pump.

The numbered equipment corresponds as follows: 1—High-pressure gas cylinders (He, N_2, CO_2, and CH_4); 2—Pressure reducing valve; 3—Six-port valve; 4, 12—Gas flowmeters; 5, 10, 14, 16—Shutoff valves; 6, 8—Gas pressure sensors; 7—Coal sample canister; 9—Thermostatic chamber; 11—Back-pressure valve; 13—Gas concentration sensor; 15—Vacuum pump; and 17—Data acquisition computer.

4.2.1.1 Gas Injection and Back-Pressure System

Figure 4.2 displays the gas injection and back-pressure system, wherein the injection subsystem comprises high-pressure gas cylinders (He, CH_4, N_2, and CO_2), a pressure reducing valve (regulation range: 0–6 MPa; accuracy: ±0.2 MPa) for controlling injection pressure, and a six-port valve (pressure rating: 0–50 MPa) for gas source selection; the back-pressure subsystem consists of a back-pressure valve (range: 0–900 Psi ≈ 0–6.2 MPa) to set the gas discharge pressure, with outflow from the coal sample canister occurring only when internal pressure exceeds the back-pressure valve's preset threshold.

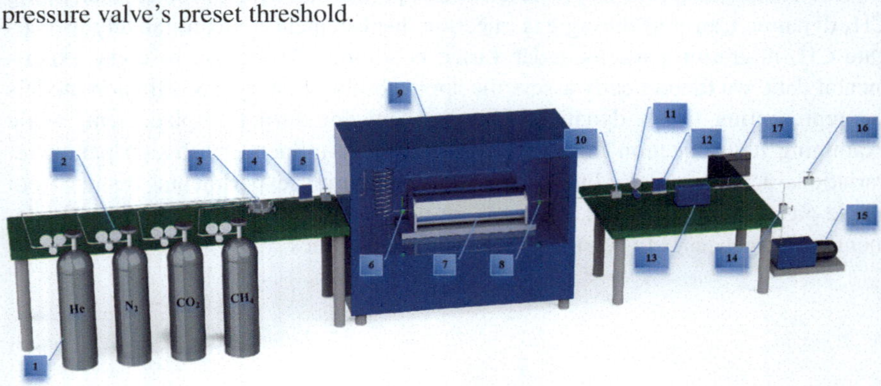

Fig. 4.1 Schematic diagram for the dynamic diffusion of CH_4 during gas injection

(a) High Pressure Gas Cylinder and Pressure Reducing Valve

(b) Six-Port Valve

(c) Back Pressure Valve

Fig. 4.2 Gas injection and back-pressure system. (**a**) High-pressure gas cylinder and pressure reducing valve. (**b**) Six-port valve. (**c**) Back-pressure valve

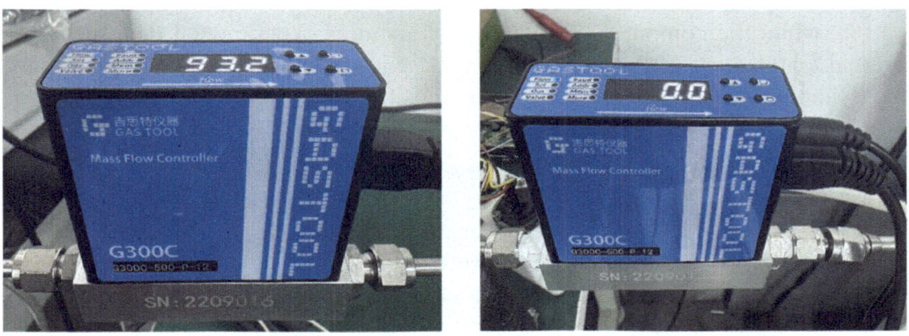

(a) Gas Injection Flow Controller

(b) Outlet Flow Monitor

Fig. 4.3 Gas injection flow control and monitoring system. (**a**) Gas injection flow controller. (**b**) Outlet flow monitor

4.2.1.2 Gas Injection Flow Control and Monitoring System

The gas injection flow control and monitoring system consists of a gas injection flow controller and an outlet flow monitor, installed at both ends of the coal sample tank. The gas injection flow controller regulates the gas injection rate, with a control range of 0–500 ml/min and a response time <100 ms. The outlet flow monitor measures the mixed gas flow rate at the exhaust port, featuring a measurement range of 0–650 ml and an accuracy of ±0.8% (Fig. 4.3).

4.2.1.3 Coal Sample Loading System

The coal sample loading system integrates a stainless steel coal sample tank and gas pressure sensors, designed for coal sample containment and experimental gas storage with an internal volume of 800 ml. The tank maintains pressure resistance up to 20 MPa and connects at both ends via gas injection steel pipes, shutoff valves, and flow control components. Gas pressure sensors monitor gas pressure at both tank interfaces, offering a measurement range of 0–25 MPa and an accuracy of ±0.1 MPa (Fig. 4.4).

4.2.1.4 Data Acquisition System

The data acquisition system integrates a continuous gas composition analyzer and a computer. The analyzer performs real-time measurement of gas volume fractions at the exhaust port, featuring a measurement range of 0–100% and an accuracy of ±0.1%, while the computer simultaneously acquires experimental data including internal temperature of the coal sample tank, gas pressure at both ends, flow rates, and dynamic variations in exhaust gas composition (Fig. 4.5).

(a) Coal Sample Tank

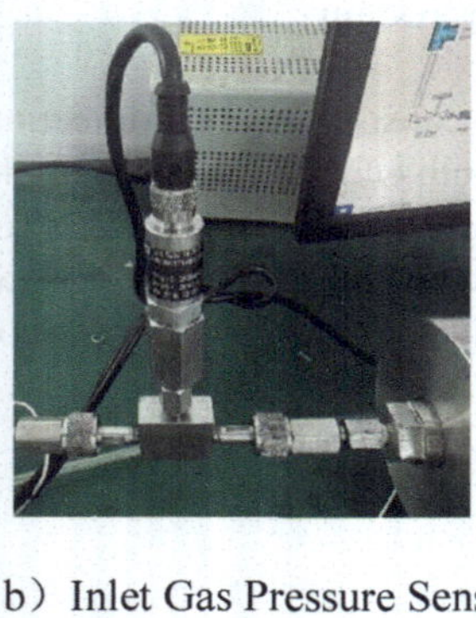

(b) Inlet Gas Pressure Sensor

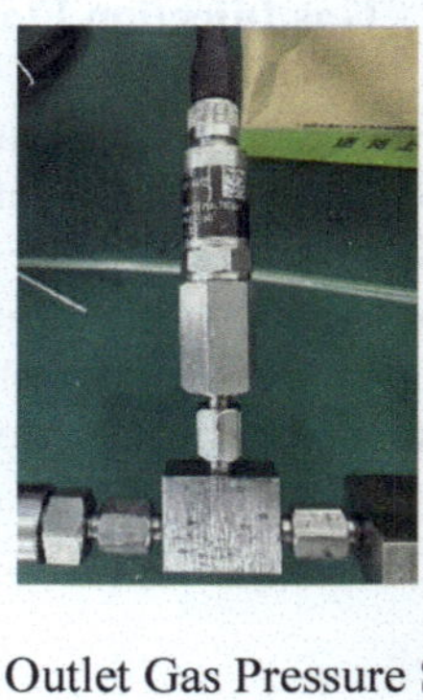

(c) Outlet Gas Pressure Sensor

Fig. 4.4 Coal sample loading system. (**a**) Coal sample tank. (**b**) Inlet gas pressure sensor. (**c**) Outlet gas pressure sensor

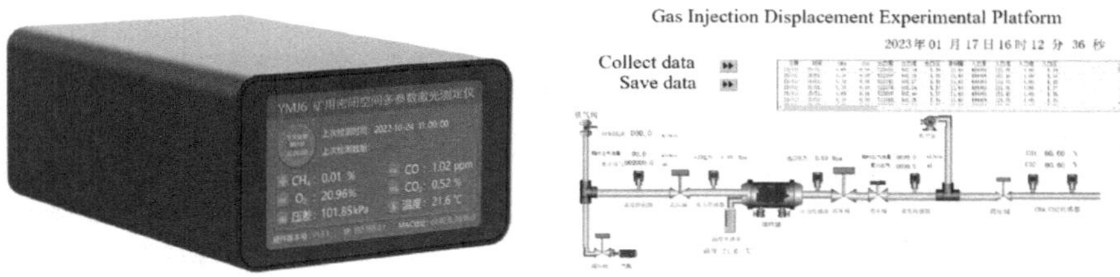

(a) Continuous Gas Composition Analyzer (b) Experimental Platform Interface

Fig. 4.5 Data acquisition system. (**a**) Continuous gas composition analyzer. (**b**) Experimental platform interface

4.2.2 Experimental Procedure

4.2.2.1 Coal Sample Preparation

The coal samples were obtained from the stable and minable 11–2# coal seam of Liuzhuang Coal Mine (Fuyang City, Anhui Province). This seam exhibits a roof lithology dominated by sandy mudstone and mudstone, with sandstone as a secondary component, and a floor lithology primarily composed of sandy mudstone and mudstone, supplemented by sandstone. Classified as a non-outburst coal seam with relatively low gas content (1.90–4.31 m^3/t), it demonstrates elevated gas emissions due to high-intensity mining activities. Fresh coal samples were sealed in fresh-keeping film for transportation to the laboratory, then processed through crushing and sieving to achieve a particle size of 1–3 mm, followed by drying for experimental use (Fig. 4.6).

4.2.2.2 Experimental Protocol and Procedures

Two experimental methodologies were conducted: single-component gas desorption and CH_4 dynamic migration under gas injection displacement, aiming to investigate CH_4 migration patterns across varying gas injection conditions.

Void Volume Calculation of Experimental System

The coal sample tank was fully loaded with prepared samples to calculate the system's void volume. Based on vessel and pipeline configuration, the void space was determined using Formula (4.1) [194], yielding a calculated value of 195.56 ml.

$$V_{void} \cdot (P_1 - P_2) = P_a \cdot V_{de} \tag{4.1}$$

where V_{viod} denotes the void volume of the experimental system; P_a denotes ambient atmospheric pressure in the laboratory; P_1 denotes the equilibrium pressure within

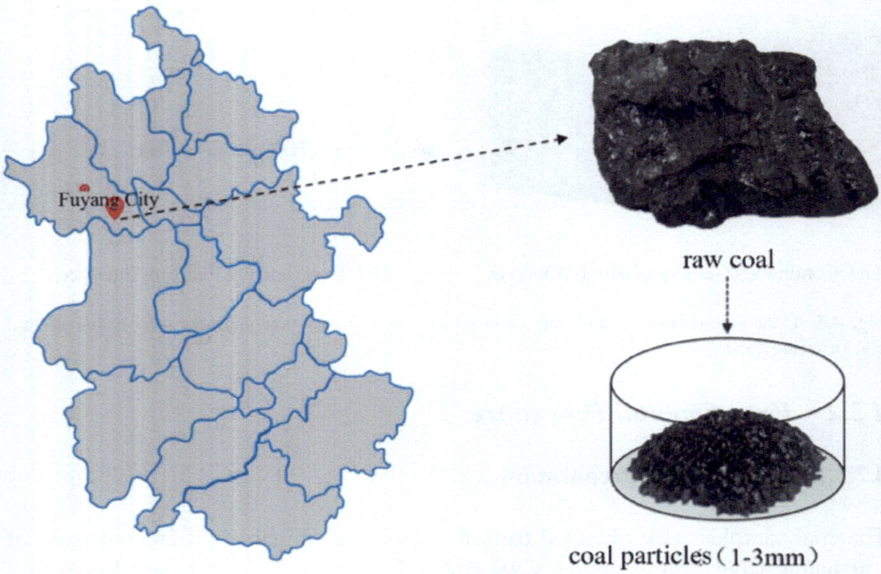

Fig. 4.6 Raw coal collection location and coal sample preparation

gas pipelines after helium injection; P_2 denotes the post-expulsion equilibrium pressure in pipelines; and V_{de} denotes the displaced helium volume during gas expulsion.

Single-Component Gas Desorption Experiment

The gas desorption experiment was conducted according to the design specifications in Table 4.1.

Experimental Procedure:

(a) Load coal sample into canister; evacuate system for ≥12 h.
(b) Close back-pressure valve; open inlet valve; inject test gas until pressure stabilizes near target adsorption equilibrium pressure; maintain for 24 hours to achieve adsorption equilibrium.
(c) Open back-pressure valve; record canister pressure and outlet instantaneous flow rate. Terminate experiment when pressure sensors and flowmeter read zero.
(d) Repeat steps a–c for single-gas (CH_4/N_2/CO_2) emission tests at adsorption equilibrium pressures of 0.5, 1.0, and 1.5 MPa.

Dynamic Transport Test for CH_4 Displacement by N_2/CO_2 Injection

Gas injection displacement extracts originally adsorbed CH_4 from coal seams by displacing it from the coal matrix. Prior to displacement, the coal system contains two primary CH_4 phases: free-phase CH_4 and adsorbed-phase CH_4. Upon gas injection, heterocomponent gases diffuse into the coal, gradually elevating pore pressure.

Table 4.1 Experimental parameters for unidirectional diffusion experiments

No.	Pre-adsorbed gas	Pre-adsorption equilibrium pressure/MPa
1	CH_4	0.5, 1.0, 1.5
2	N_2	0.5, 1.0, 1.5
3	CO_2	0.5, 1.0, 1.5

Outflow commences only when the internal pore pressure exceeds the original adsorption equilibrium pressure—a critical threshold neglected in prior studies where outlets were erroneously set to atmospheric pressure [195–197]. Based on this physiochemical principle, we designed a field-realistic experimental protocol as follows:

(a) System Integrity Check and Coal Loading.
Open the inlet valve to pressurize both inlet and outlet to the target pre-adsorption pressure. Close the inlet valve and set the back-pressure valve to the pre-adsorption pressure. Maintain this state; if inlet/outlet pressures remain unchanged for 12 hours, confirm system integrity. Fill the adsorption canister with sieved and dried coal particles.

(b) Vacuum Degassing.
Close the inlet valve, open the outlet valve, and deactivate the back-pressure function. Evacuate coal particles via vacuum pump for 1 h.

(c) Adsorption Equilibrium.
Open the CH_4 cylinder and inlet valve to gradually pressurize the canister and gas lines to the target pressure. Set the back-pressure valve to the pre-adsorption pressure. Maintain conditions until adsorption equilibrium is achieved (confirmed by stable canister pressure for 8 consecutive hours).

(d) Gas Injection Displacement.
Close the inlet valve and open the purge outlet of the six-port valve to evacuate residual CH_4 gas from the pipeline between the inlet valve and gas cylinder; open the gas cylinder and inlet valve; program the injection flow control sequence via software–hardware integration of the mass flow controller, setting the flow injection mode for the inlet mass flow controller to initiate a 4-h gas injection displacement experiment, with results automatically recorded and saved in the software.

(e) Post-Injection Observation.
Through software-mass flow controller communication, program the system to terminate gas injection after 4 h; following injection cessation, open the back-pressure valve to discharge gases from the adsorption canister.

Repeat steps a–e for subsequent gas injection displacement tests. The experimental design for dynamic transport characterization during N_2/CO_2 displacement of CH_4 is detailed in Table 4.2.

To investigate the dynamic transport behavior of CH_4 in coal seams under flow-rate disturbance conditions, a baseline average flow rate of 100 ml/min was established. Two periodic variable flow injection modes (sinusoidal variation and stepped variation) with six amplitude levels were implemented, alongside one conventional constant-flow injection mode for comparison. Detailed gas injection flow sequences are specified in Table 4.2, with schematic representations of periodic flow variations provided in Fig. 4.7.

Table 4.2 Experimental program design for the determination of dynamic transport properties of N_2, CO_2, driven CH_4

No.	Injected gas	Pre-adsorbed gas	Gas injection method	Gas injection flow rate/(ml/min)	Pre-adsorption equilibrium pressure/MPa
1	N_2	CH_4	Constant flow	$y = 100$	1.0
			Constant flow	$y = 200$	
			Constant flow	$y = 300$	
			Sinusoidally variable flow	$y = (100/75/50)\sin\frac{\pi}{5}t + 100$	1.0
			Stepwise variable flow	$y = \begin{cases} 200/175/150, & 0+10n \le t \le 5+10n \\ 0/25/50, & 5+10n \le t \le 10+10n \end{cases} (n = 0,1,2,3\ldots)$	
2	CO_2	CH_4	Constant flow	$y = 100$	1.0
			Constant flow	$y = 200$	
			Constant flow	$y = 300$	

Note: In the table, t denotes injection time, and y represents the injection flow rate at time t

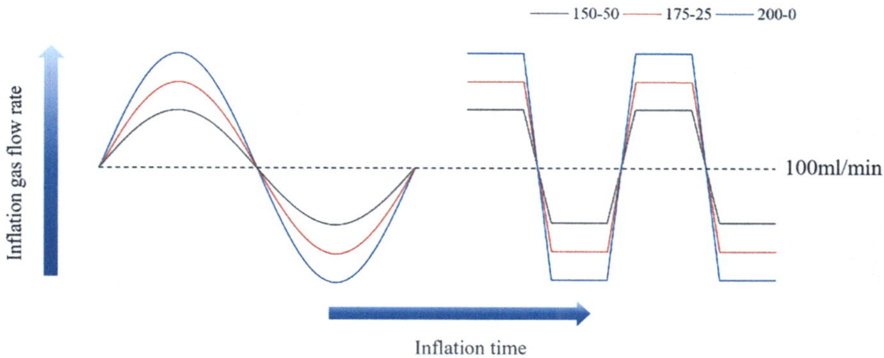

Fig. 4.7 Variation of injected gas flow rate

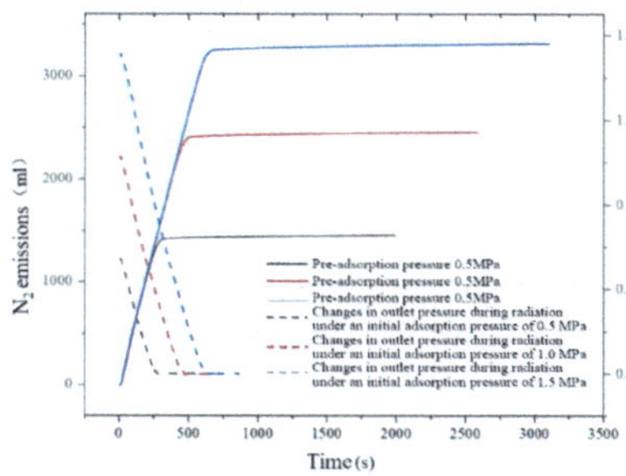

(a) N2 single gas

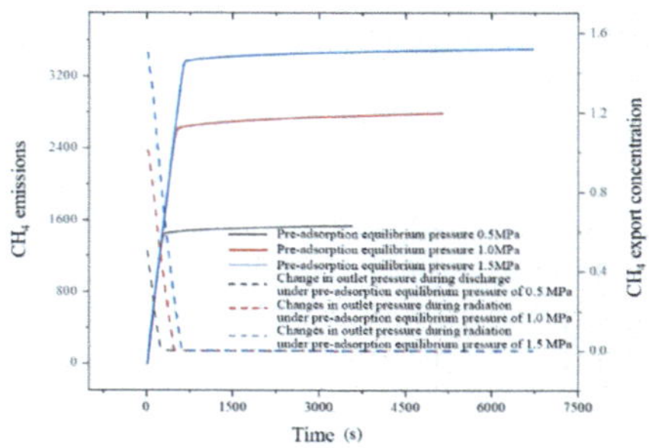

(b) CH4 single gas

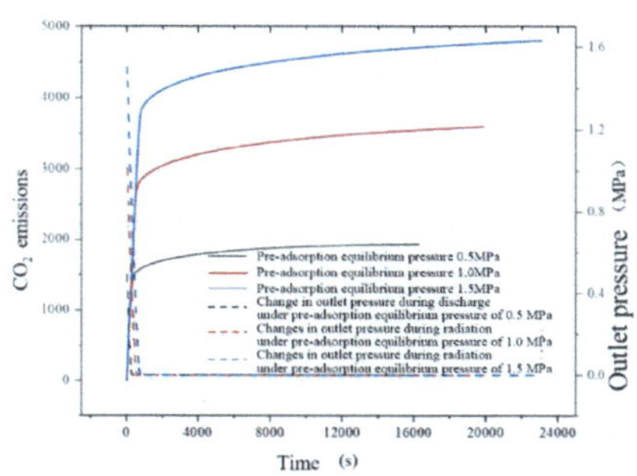

(c) CO2 single gas

Fig. 4.8 Unidirectional diffusion emissions of N_2, CO_2, and CH_4. (**a**) N_2 single gas. (**b**) CH_4 single gas. (**c**) CO_2 single gas

4.3 Single-Gas Emission Experiments in Coal

Figure 4.8 presents experimental results for N_2, CH_4, and CO_2 emissions under adsorption equilibrium pressures of 0.5, 1.0, and 1.5 MPa. The emission process for all gases and pressure levels exhibits two distinct stages: In stage I (early diffusion),

Table 4.3 Experimental results of N_2, CO_2, and CH_4 diffusion

Gas species	Pre-adsorption equilibrium pressure (MPa)	Depressurization duration to outlet pressure zero(s)	Total gas emission volume(ml)	Total emission time(s)
N_2	0.5	232	1457.28	1990
	1.0	463	2463.15	2576
	1.5	631	3317.17	3104
CH_4	0.5	254	1536.28	3574
	1.0	484	2792.92	5150
	1.5	617	3501.73	6716
CO_2	0.5	280	1934.62	16,296
	1.0	536	3600.51	19,899
	1.5	741	4802.22	23,088

disruption of adsorption equilibrium rapidly reduces the pressure gradient between the coal sample canister and atmosphere, expelling free-phase gas while triggering substantial desorption of adsorbed gas; during stage II (late diffusion), the pressure gradient progressively reduces to zero, with emissions dominated by adsorbed-phase gas desorption. Notably, at identical equilibrium pressures, cumulative emissions consistently follow $CO_2 > CH_4 > N_2$—a hierarchy governed by coal's gas adsorption affinity ($CO_2 > CH_4 > N_2$), aligning with established literature. Stronger adsorption enables coal to retain more gas at equilibrium; consequently, greater volumes of highly adsorbable CO_2 are ultimately emitted compared to CH_4 and N_2. Table 4.3 quantifies key results: for any given gas, increased adsorption equilibrium pressure elevates total emissions, while the duration for the canister-atmosphere pressure gradient to reach zero correlates positively with coal's adsorption affinity for that gas.

4.4 Macroscopic Dynamics of CH_4 Transport During Gas Injection Displacement

4.4.1 Impact of Injection Flow Rate on CH_4 Transport Behavior

Using N_2 injection for coalbed CH_4 displacement as a case study, this section examines flow-rate effects on macroscopic CH_4 transport dynamics.

4.4.1.1 Temporal Evolution of Outlet Gas Composition

Figure 4.9 depicts the variation in outlet gas composition with displacement time at injection flow rates of 100, 200, and 300 ml/min under a constant CH_4 pre-adsorption pressure of 1.0 MPa. Two distinct temporal phases characterize the compositional shifts across all flow conditions.

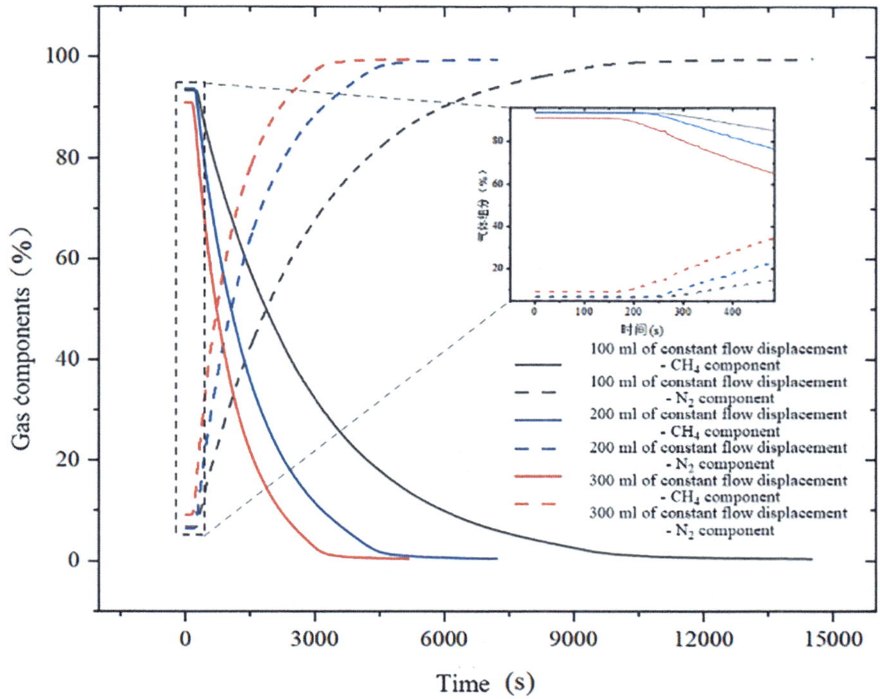

Fig. 4.9 Variation of each gas component in the process of N_2 injection to drive coal grain CH_4 under each gas injection flow condition

Table 4.4 Time elapsed for N_2 injection to drive out coal grain CH_4 process I and II phases

Phase	Phase I(s)	Phase II(s)
Constant flow 100 ml	259	13,400
Constant flow 200 ml	193	7021
Constant flow 300 ml	156	4949

Phase I: During initial N_2 injection, the injected gas primarily displaces free-phase CH_4, resulting in CH_4-dominated outlet composition where its concentration remains relatively constant. Phase II: Continuous N_2 infusion diffuses into coal matrix micropores, simultaneously promoting CH_4 transport and triggering desorption through partial pressure effects. However, CH_4 desorption rates are exceeded by N_2 injection volumes. Consequently, N_2 concentration rises due to dilution effects while CH_4 concentration declines. By late Phase II, N_2-driven CH_4 desorption-diffusion ceases as N_2 becomes the dominant gas phase; CH_4 desorption terminates, stabilizing both components' concentrations. Table 4.4 quantifies phase durations across flow rates, revealing: Higher injection rates reduce both phase durations—attributed to amplified CH_4 concentration gradients between matrix pores and macropores. Elevated N_2 flux accelerates adsorbed CH_4 desorption-diffusion kinetics, thereby shortening each phase.

4.4.1.2 Evolution of Outlet Gas Production Rate and Pressure Gradient

Figures 4.10 and 4.11, respectively, depict temporal variations in outlet gas production rate and pressure gradient during N_2-driven CH_4 displacement across injection flow rates (100, 200, and 300 ml/min).

Figure 4.10 reveals that under constant-flow N_2 injection, the outlet mixed gas production rate exhibits a consistent three-phase growth pattern (rapid increase → gradual rise → stabilization) across all flow rates as displacement progresses. This trend precisely mirrors the pressure gradient evolution between the coal sample canister and outlet (Fig. 4.11), which similarly progresses through rapid increase → gradual rise → stabilization.

Notably, under all three injection flow rates (100, 200, and 300 ml/min), a transient deceleration in outlet flow growth occurs during initial injection—particularly pronounced at 100 ml/min, where a slight decline is observed. This phenomenon arises because early-stage N_2 diffusion into interparticle spaces and matrix micropores mobilizes high-concentration CH_4 toward the outlet. However, low initial N_2

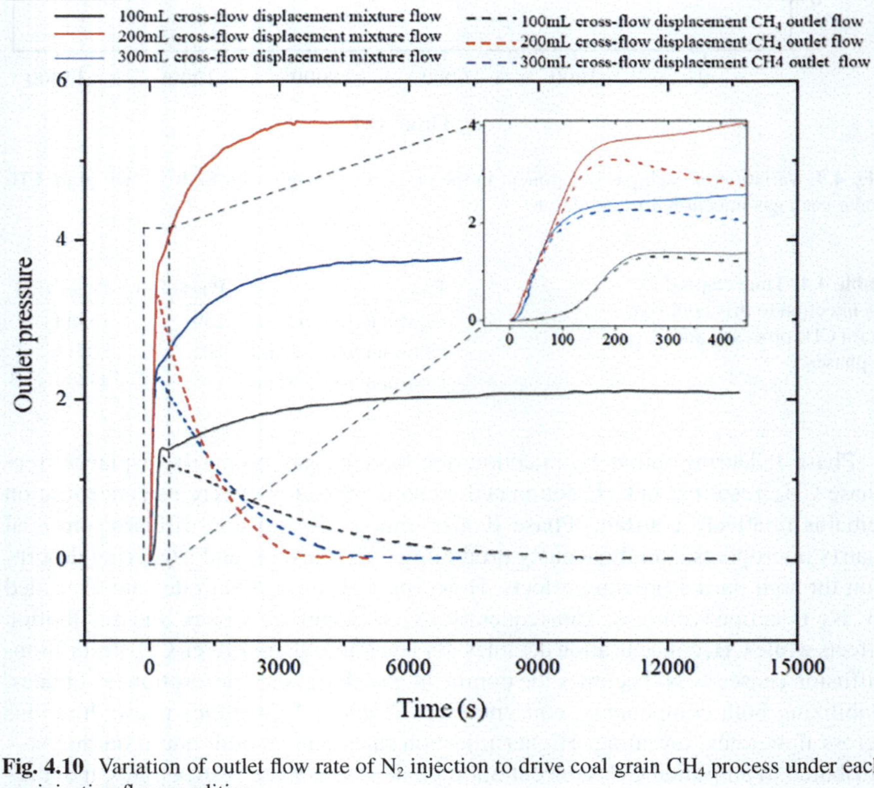

Fig. 4.10 Variation of outlet flow rate of N_2 injection to drive coal grain CH_4 process under each gas injection flow condition

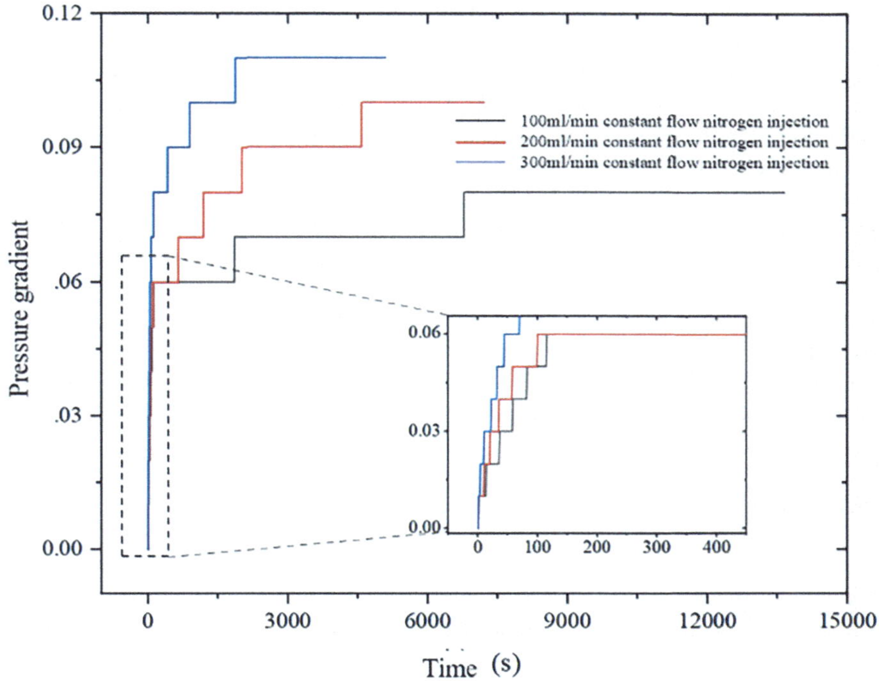

Fig. 4.11 Variation of coal sample tank and outlet gas pressure gradient in the process of injecting N_2 to drive CH_4 from coal particles under each gas injection flow condition

concentration within the coal results in minimal concentration gradients between matrix pores and macropores, slowing CH_4/N_2 diffusion and limiting outlet flow. As N_2 injection continues, cumulative N_2 influx increases coal N_2 concentration. Following N_2 breakthrough [198], free-phase CH_4 in fractures depletes rapidly, and N_2 opens existing micro-fracture networks, amplifying CH_4 concentration gradients and accelerating migration. Consequently, outlet flow rises until outflow exceeds injection rate, temporarily depleting internal gas reserves and reducing concentration gradients. This causes a brief outlet flow deceleration or decline. Sustained N_2 injection replenishes coal gas reservoirs, allowing outlet flow to resume its ascent. Ultimately, outlet flow stabilizes when injection-to-production equilibrium establishes a constant ratio and stable concentration gradients.

Figure 4.10 further delineates the evolution of CH_4 outlet flow rate across injection flow rates, revealing a triphasic pattern: Stage I (Rapid Rise): N_2 injection elevates the canister-outlet pressure gradient, driving rapid expulsion of free-phase CH_4 that dominates outlet flux (>85%), synchronizing CH_4 flow with pressure surge. Stage II (Progressive Decline): Continuous N_2 injection depletes free CH_4 while establishing CH_4 concentration gradients across matrix pores. Adsorbed CH_4 desorbs as N_2 diffuses inward, but declining concentration gradients progressively diminish desorption rates. Despite rising total outlet flow, CH_4-specific flow decreases due to reduced desorbed volumes per unit time, making desorbed CH_4 the dominant outlet source. Stage III (Stabilization): Near-complete adsorbed CH_4 depletion stabilizes CH_4 flow at low levels: 0.011 ml/s (100 ml/min), 0.020 ml/s (200 ml/min), and 0.029 ml/s (300 ml/min), reflecting flow-rate-dependent residual emissions.

Then, Fig. 4.10 further demonstrates that increased injection flow rates elevate both the stabilized and peak outlet mixed gas flow rates: stabilized flows reach approximately 2.08, 3.76, and 5.50 ml/s, while peak flows achieve 1.30, 2.27, and 3.29 ml/s at 100, 200, and 300 ml/min injection rates, respectively. This positive correlation arises because higher N_2 influx accelerates and intensifies the pressure gradient between the coal canister and outlet (Fig. 4.11), thereby amplifying mixed gas and CH_4 flow rates while shortening the time to peak values.

In summary, elevated injection flow rates accelerate CH_4 transport dynamics during gas displacement by enhancing advective forces and pressure-driven mobilization.

4.4.1.3 Evolution of Cumulative CH_4 Production

Figure 4.12 displays cumulative CH_4 production and total experimental duration during N_2 displacement of CH_4 across injection flow rates. For all flow rates, cumulative CH_4 production exhibits a rapid initial increase followed by gradual growth

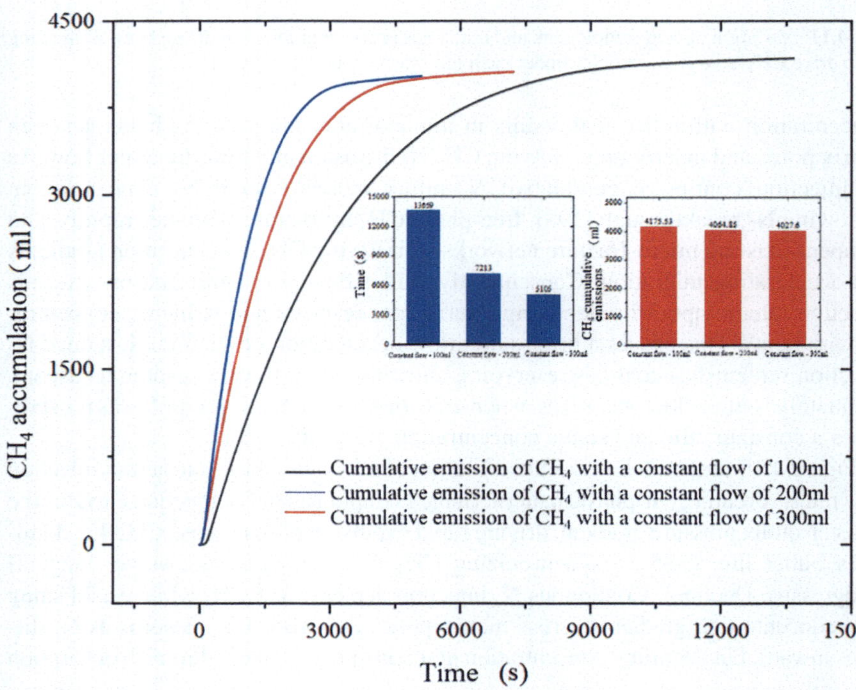

Fig. 4.12 Cumulative emission changes of N_2 injection to replace CH_4 in coal particles under each gas injection flow rate and the experimental time

with displacement time. Higher injection rates significantly reduce total experimental duration: at 100, 200, and 300 ml/min, cumulative CH_4 production totals 4075.53, 4064.85, and 4027.60 ml, respectively—demonstrating comparable ultimate yields (mean variation <1.2%) and confirming minimal flow-rate impact on total recoverable CH_4. Conversely, displacement durations (13,659, 7213, and 5105 s) show an inverse correlation with injection flow rate. Thus, while elevated flow rates accelerate process completion, they exert negligible influence on cumulative CH_4 production—consistent with findings in Sect. 3.4.2 of this book.

4.4.2 Impact of Injection Gas Type on CH_4 Transport Behavior

4.4.2.1 Temporal Evolution of Outlet Gas Composition

Figure 4.13 presents the temporal evolution of CH_4 concentration profiles during CO_2/N_2 displacement of CH_4 across injection flow rates.

Observations from Fig. 4.13 indicate that constant-flow CO_2 injection exhibits a CH_4 concentration evolution pattern similar to N_2 across both Stage I and Stage II (durations quantified in Table 4.5), yet critical distinctions emerge between the two injectants:

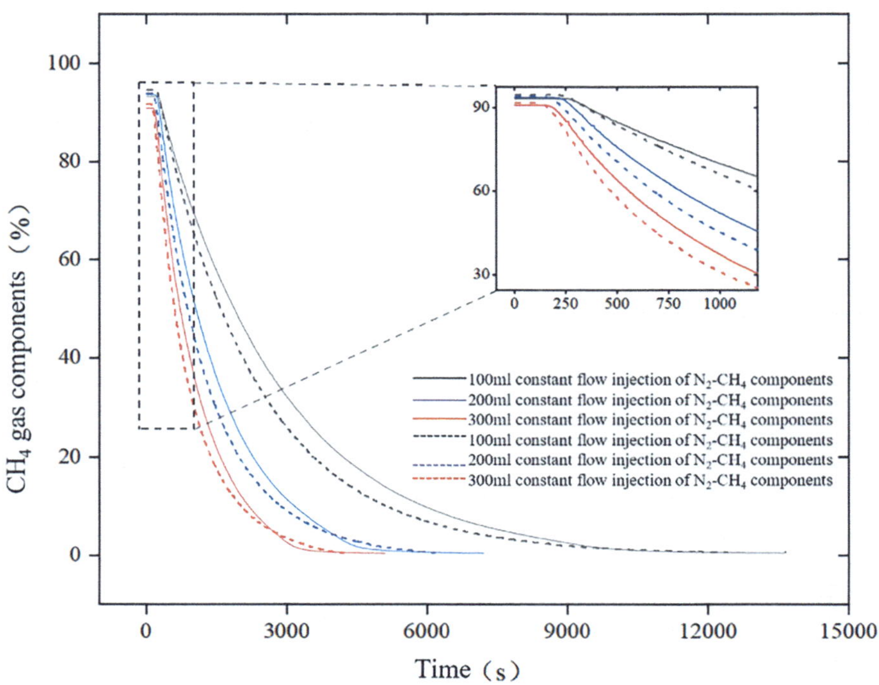

Fig. 4.13 Changes of CH_4 gas fraction in the process of injecting CO_2/N_2 to drive CH_4 from coal under each gas injection flow condition

Table 4.5 Time elapsed during Phase I and II of CO_2 injection to drive out coal grain CH_4 process

Phase	Phase I(s)	Phase II(s)
Constant flow 100 ml/min	208	12,201
Constant flow 200 ml/min	150	6070
Constant flow 300 ml/min	123	4155

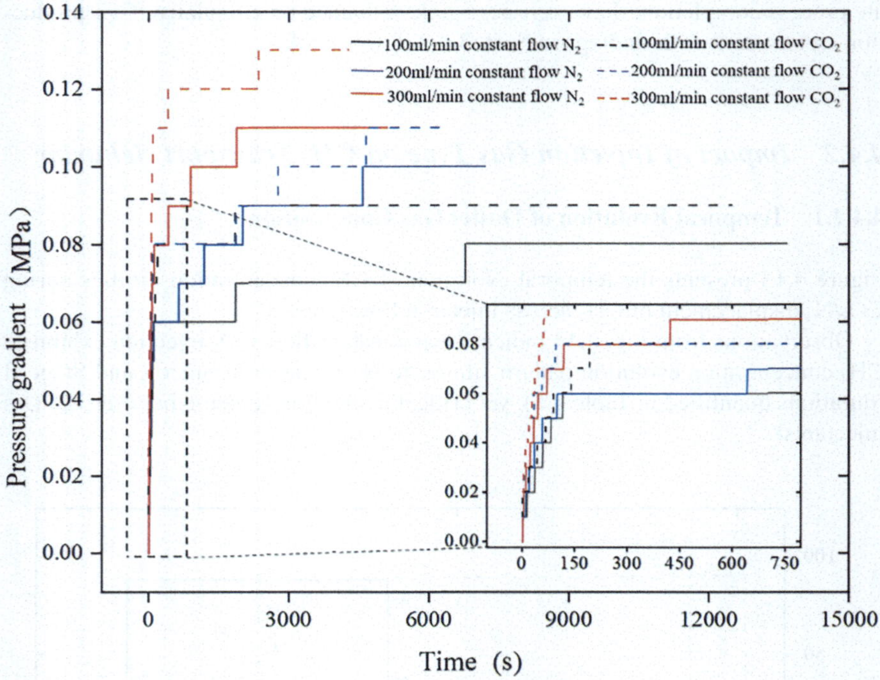

Fig. 4.14 Variation of coal sample tank and outlet gas pressure gradient in the process of injecting CO_2/N_2 to drive CH_4 from coal under each gas injection flow condition

Phase I: Under CO_2 injection flow rates of 100, 200, and 300 ml/min, the durations recorded were 208, 150, and 123 s, respectively—all shorter than those observed when N_2 was used as the injection gas source during Phase I. This phenomenon can be explained by Fig. 4.14. Due to the dual effects of competitive adsorption and partial pressure, CO_2 injection enables the desorption of more CH_4 gas per unit time compared to N_2 injection. This leads to a faster increase in the gas pressure gradient within the coal sample canister and at the outlet, resulting in accelerated gas flow velocity and earlier detection of CO_2 at the outlet. This finding differs from results obtained in experiments involving CO_2 injection to displace CH_4 from coal blocks [199–201]. The discrepancy arises because coal blocks are inherently denser than coal particle aggregates, and the adsorption-induced swelling caused by injected CO_2 is insufficient to affect gas migration within the coal matrix. Instead, the permeability of the coal sample emerges as the key factor influencing gas breakthrough time.

Phase II: As the displacement process progresses, a significant amount of CO_2 is injected into the coal sample canister. By this stage, most of the original free-state CH_4 has been expelled through the "flushing effect" of Phase I. Consequently, the CH_4 detected at the outlet during Phase II primarily originates from the desorption of originally adsorbed CH_4. As shown in Fig. 4.13, during Phase II under various injection flow rates, the CH_4 concentration at the outlet when using CO_2 as the gas source first falls below and then exceeds that observed when using N_2. In other words, compared to N_2 injection, the CH_4 concentration at the outlet declines more rapidly initially but slows down later when CO_2 is used as the injection gas. According to literature [202], the desorption curve of CH_4 in coal can be broadly divided into a "rapid desorption-diffusion zone" and a "slow desorption-diffusion zone" based on desorption time. This zoning phenomenon is attributed to the diffusion pathways and resistance of CH_4 within the coal matrix. During CH_4 desorption in coal, the adsorbed CH_4 on the coal matrix surface desorbs and diffuses first, followed by adsorbed CH_4 in relatively larger internal pores, and finally adsorbed CH_4 in smaller internal pores. The first category can be termed "easily desorbed" CH_4, while the latter two constitute "difficult-to-desorb" CH_4. Given the identical pre-adsorption CH_4 gas pressure (1 MPa) across all experiments, the quantities of CH_4 adsorbed on the coal matrix surface, larger pores, and smaller pores are consistent under each experimental condition. Therefore, during the displacement process, adsorbed CH_4 on the coal matrix surface undergoes desorption first. Due to the dual effects of CO_2's competitive adsorption and partial pressure, CH_4 desorbs faster than when N_2 is used as the injection gas source. This results in a more rapid decline in CH_4 concentration at the outlet during the early Phase II when CO_2 is selected as the injection gas. In the later Phase II, owing to the difference in coal's adsorption affinity for the two gases, CO_2 penetrates the coal matrix micropores more readily than N_2. It competitively displaces CH_4 adsorbed in smaller pores, promoting desorption and diffusion of adsorbed CH_4 from pore surfaces into free spaces. Consequently, the CH_4 concentration at the outlet becomes marginally higher during CO_2 injection displacement compared to N_2 injection displacement.

The above experimental phenomena are consistent with the description in Sect. 3.4.1.

4.4.2.2 Variation in Outlet Gas Production Flow Rate

Figure 4.15 illustrates the variation curves of the outlet mixed gas flow rate during the displacement of CH_4 from coal by injecting CO_2/N_2 under different gas injection flow rates.

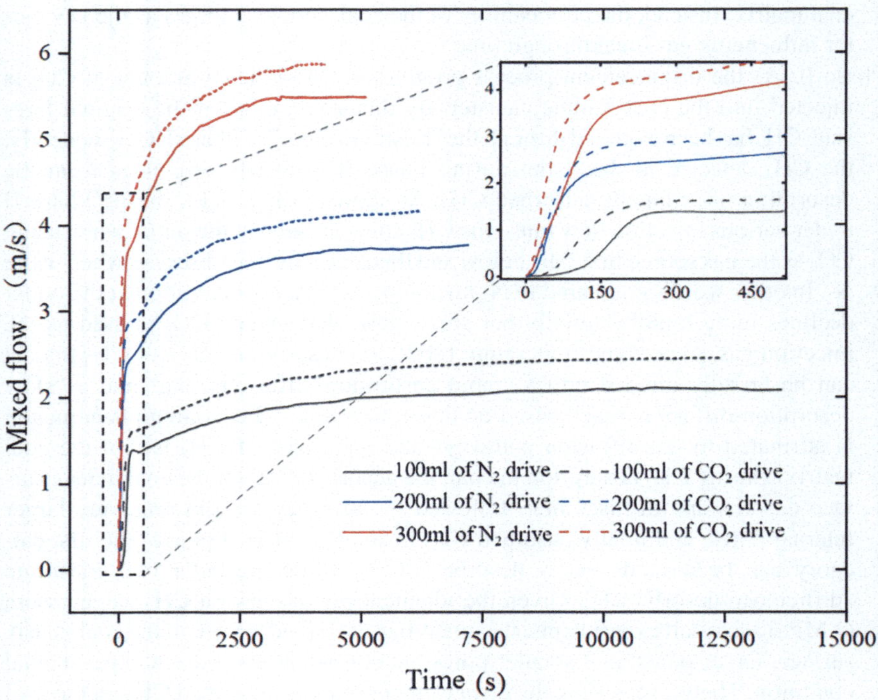

Fig. 4.15 Variation of gas mixture flow rate at the outlet of CO_2/N_2 injection for the process of CH_4 replacement of coal particles under each gas injection flow rate condition

It can be observed that, consistent with the conditions under N_2 injection displacement, the process of displacing CH_4 from coal using CO_2 also exhibits a trend of rapid increase → slow increase → gradual stabilization. Under identical gas injection flow rates, compared to N_2 as the gas source, the use of CO_2 results in higher outlet mixed gas flow rates and a faster initial rise. The peak mixed gas flow rates stabilize at approximately 2.45, 4.16, and 5.89 ml/s, respectively. During CO_2 injection for CH_4 displacement in coal, the desorption of adsorbed CH_4 is promoted through two mechanisms: competitive adsorption (where CO_2 occupies the adsorption sites originally held by CH_4) and reduction of CH_4 partial pressure. In contrast, N_2 injection promotes CH_4 desorption solely through partial pressure reduction. Consequently, compared to N_2 injection displacement, CO_2 injection facilitates a greater volume of originally adsorbed CH_4 desorption per unit time. Moreover, the coal's CO_2 adsorption capacity per unit time is significantly lower than the injection flow rate. These factors account for the phenomena observed in Fig. 4.15. The gas pressure gradient curves within the coal sample canister and at the outlet for both gas sources, as shown in Fig. 4.14, further corroborate this mechanism.

Figure 4.16 presents the variation curves of CH_4 outlet flow rate during gas injection displacement under different injection flow rates. It can be observed that the

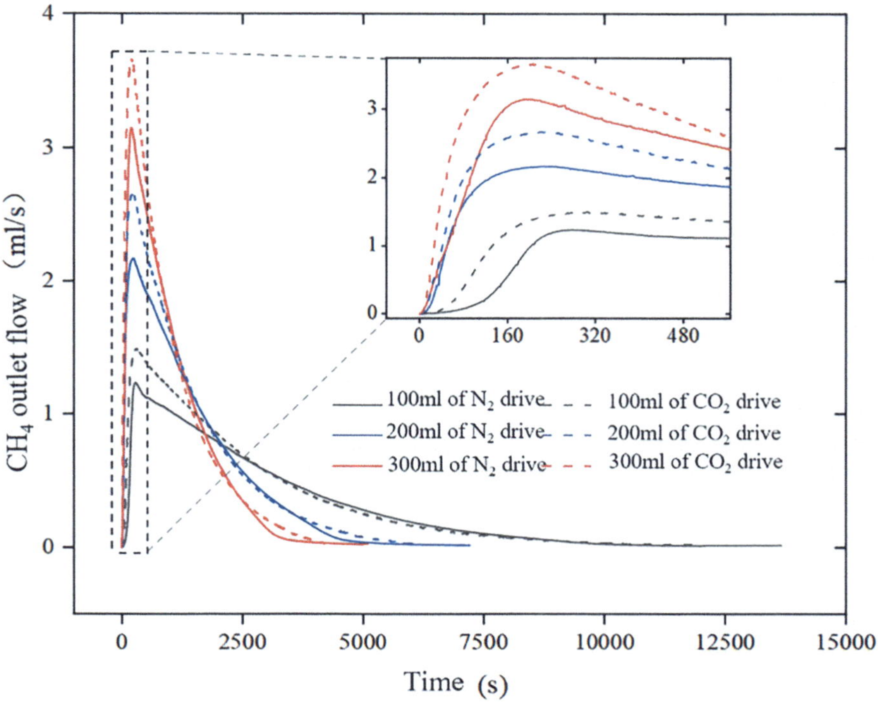

Fig. 4.16 Variation of CH_4 gas flow rate at the outlet of CO_2/N_2 injection process for CH_4 replacement of coal particles under each gas injection flow condition

trend of CH_4 outlet flow rate is consistent for both gas sources during displacement. During CO_2 injection displacement, the CH_4 outlet flow variation also exhibits three stages, though with slight differences: Stage I: Compared to N_2 injection displacement, CH_4 outlet flow rises faster and reaches higher peak values during CO_2 injection displacement. Under injection flow rates of 100, 200, and 300 ml/min, the maximum CH_4 outlet flows are 1.49, 2.68, and 3.83 ml/s, respectively. Stage II: During this phase, the CH_4 outlet flow declines more rapidly under CO_2 injection than under N_2 injection. This occurs because CO_2's dual mechanisms—competitive adsorption and partial pressure reduction—accelerate the desorption of "easily desorbed" CH_4. Stage III: As injection continues, "difficult-to-desorb" CH_4 within the coal matrix begins to desorb and diffuse into free spaces under CO_2 injection, replenishing the outlet CH_4 flow. Consequently, the CH_4 outlet flow in this stage exceeds that observed during N_2 injection displacement. Ultimately, under injection flow rates of 100, 200, and 300 ml/min, the CH_4 outlet flow stabilizes at approximately 0.012, 0.021, and 0.029 ml/s, respectively—aligning closely with the corresponding values observed during N_2 injection displacement.

4.4.2.3 Variation in CH_4 Cumulative Emission

Figure 4.17 illustrates the variation in cumulative CH_4 emissions during the displacement of CH_4 from coal by injecting CO_2/N_2 under various injection flow rates. As observed, under all flow conditions, the cumulative CH_4 emission curve during CO_2 injection exhibits a trend of rapid initial increase followed by a slower increase—consistent with that observed during N_2 injection displacement. The total cumulative CH_4 emissions under injection flow rates of 100, 200, and 300 ml/min are 4203.66, 4252.16, and 4195.81 ml, respectively. These values are all higher than those recorded during N_2 injection displacement under equivalent flow rates, aligning with the conclusion drawn in Sect. 3.4.1 regarding the influence of gas source type on cumulative CH_4 emissions. However, while an increase in injection flow rate accelerates the rate of cumulative CH_4 emission growth for the same gas source, the total cumulative emissions show insignificant differences. Figure 4.18 presents the experimental durations for CO_2/N_2 injection displacement under each flow rate. The total durations for CO_2 injection at 100, 200, and 300 ml/min are 12,409, 6220, and 4278 s, respectively—all shorter than those for N_2 injection experiments under equivalent flow rates. This indicates that CO_2 injection accelerates CH_4 migration in coal compared to N_2 injection.

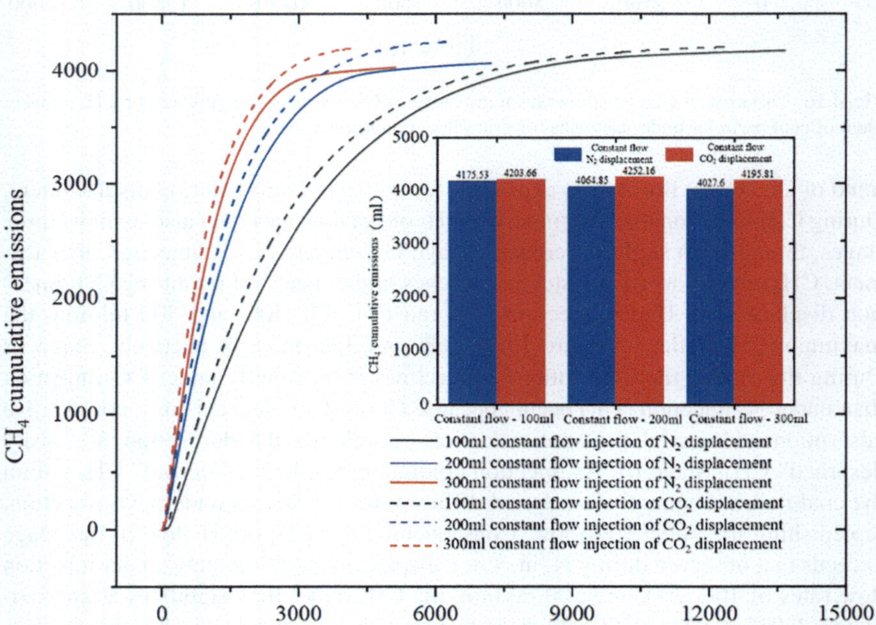

Fig. 4.17 Changes in cumulative emissions from the injection of CO_2/N_2 to replace coal grain CH_4 process under each gas injection flow condition

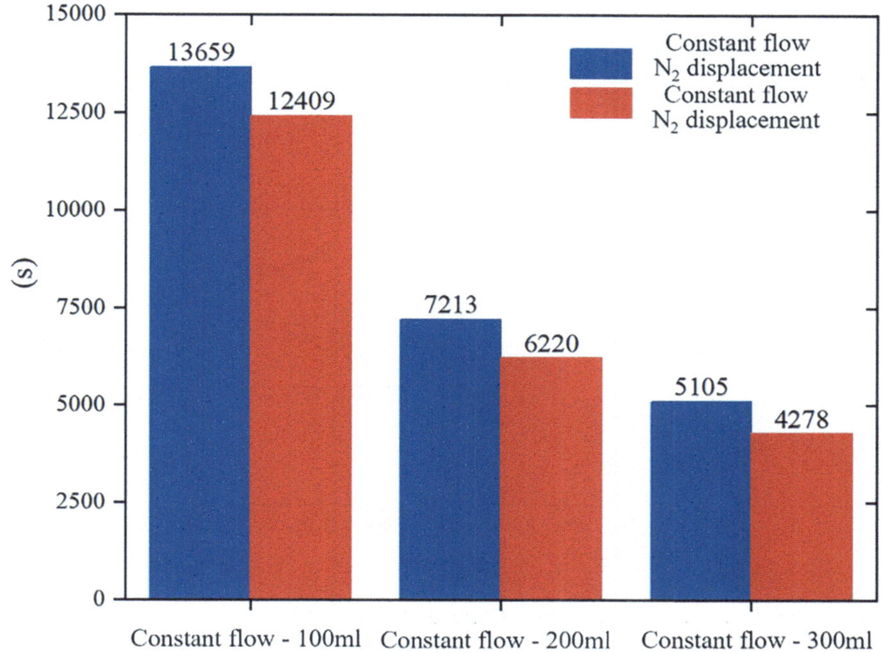

Fig. 4.18 Experimental time spent on injecting CO_2/N_2 to drive coal grain CH_4 process under each gas injection flow condition

4.4.3 Influence of Gas Injection Methods on Dynamic CH_4 Migration Patterns in Coal

This section employs N_2 as the injection gas source to investigate the impact of different gas injection methods on CH_4 displacement in coal. Based on the principle of equivalent injection duration and volume, all injection methods maintained an average flow rate of 100 ml/min. Experimental results are compared with those from the constant-flow N_2 injection method at 100 ml/min. The total injection duration was 4 h, with a cumulative injection volume of 24,000 ml.

4.4.3.1 Variation in Outlet Gas Composition

Figure 4.19 illustrates the variation in outlet gas composition with displacement time under different gas injection methods. Table 4.6 presents the N_2 breakthrough times for each injection method. The data shows that under sinusoidal injection modes (200 → 0, 175 → 25, and 150 → 50 ml/min), N_2 breakthrough times are 211, 229, and 231 s, respectively. For stepwise injection modes (200 → 0, 175 → 25, and 150 → 50 ml/min), breakthrough times are 192, 203, and 212 s, respectively. In

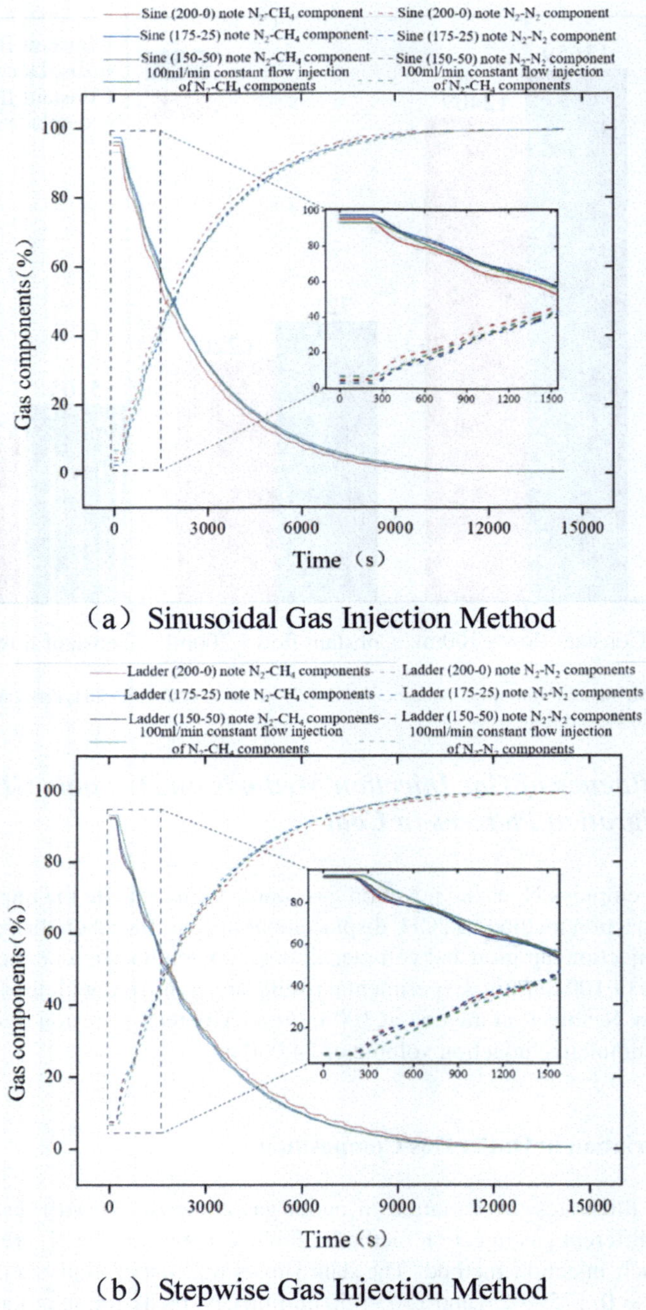

Fig. 4.19 Changes in volume fraction of gas components at the outlet end under the conditions of different gas injection methods. (**a**) Sinusoidal gas injection method. (**b**) Stepwise gas injection method

Table 4.6 N_2 breakthrough time for different gas injection methods

Gas injection method	Breakthrough time (s)
Sinusoidal (200–0 ml/min)	211
Sinusoidal (175–25 ml/min)	229
Sinusoidal (150–50 ml/min)	231
Stepwise (200–0 ml/min)	192
Stepwise (175–25 ml/min)	203
Stepwise (150–50 ml/min)	212
Constant flow 100 ml/min	259

contrast, constant-flow injection at 100 ml/min yields an N_2 breakthrough time of 259 s. All cyclic variable-rate injection methods exhibit earlier breakthrough times compared to constant-flow injection. This occurs because during the initial half-cycle of cyclic injection, the N_2 flow rate exceeds that of constant-flow injection, creating steeper gas concentration gradients in the coal matrix and micropore system. Consequently, the pressure gradient between the coal sample canister interior and the outlet intensifies, accelerating CH_4 migration and shortening N_2 breakthrough time. Notably, stepwise injection maintains the highest initial flow rate for the longest duration during its first half-cycle compared to both constant-flow and sinusoidal methods, resulting in the most rapid N_2 breakthrough.

Under constant-flow injection at 100 ml/min, the CH_4 concentration exhibits a continuous downward trend over time, while the N_2 concentration rises steadily. In contrast, both sinusoidal and stepwise injection methods produce fluctuating downward trends in CH_4 concentration and fluctuating upward trends in N_2 concentration. Specifically, under cyclic variable-rate injection conditions after N_2 breakthrough, CH_4 concentration decreases rapidly during high-flow injection phases but declines more slowly during low-flow injection phases. Consequently, CH_4 concentration exhibits periodic fluctuations corresponding to injection flow rate changes. This occurs primarily because: During slower-decline phases (low-flow injection), adsorbed CH_4 in the coal matrix and micropores continues desorption-diffusion processes. However, the low N_2 injection flow provides insufficient driving force to mobilize CH_4 migration within the coal body, resulting in increased CH_4 retention. When N_2 injection flow increases, the higher-flow N_2 stream flushes the retained CH_4 toward the outlet. This temporarily slows the observed downward trend of CH_4 concentration at the outlet. Conversely, during high-flow injection phases, increased N_2 volume per unit time enhances dilution effects, accelerating the decline rate of CH_4 concentration. Collectively, these mechanisms produce the characteristic fluctuating downward trend in outlet CH_4 concentration under cyclic variable-rate injection conditions.

It should be noted that this cyclical "rapid-slow" decline pattern in CH_4 concentration is more pronounced under stepwise injection conditions. The primary reason is that in sinusoidal injection, the gas flow rate undergoes continuous periodic variation over time, never stabilizing at maximum or minimum values. To elaborate: In sinusoidal injection, when the gas flow rate reaches its minimum, it immediately begins to increase over time, providing the energy required to mobilize gas migration within the coal within a shorter timeframe. Conversely, in stepwise injection, the gas flow rate remains at its minimum level for an extended duration. Thus, during low N_2 injection flow phases, stepwise injection causes greater retention of free-state CH_4 within the coal body for longer periods. When the injection flow increases, the outlet CH_4 concentration declines more slowly and may even exhibit a brief slight increase during later displacement stages.

4.4.3.2 Variation in Outlet Gas Production Flow Rate

Figure 4.20 presents the temporal variations of mixed gas flow rate and CH_4 flow rate at the outlet during displacement under different injection methods. The results are compared with constant-flow injection at 100 ml/min based on the principle of equivalent duration and injection volume.

Figure 4.20 further reveals that during flow variations between 200 and 0 ml/min, stepwise injection produces higher peak values but lower minimum values for both mixed gas and CH_4 flow rates at the outlet compared to sinusoidal injection. Concretely, during high-injection-flow phases, stepwise injection yields peak outlet mixed gas and CH_4 flow rates of 3.79 and 2.38 ml/s, respectively, while sinusoidal injection achieves 3.45 and 2.24 ml/s. Conversely, during low-injection-flow transitions (high-low-high), stepwise injection produces minimum mixed gas and CH_4 flows of 0.28 and 0.0028 ml/s, respectively, compared to sinusoidal injection's 0.42 and 0.0042 ml/s. This occurs because stepwise injection with instantaneous switching between 200 and 0 ml/min creates abrupt, sustained pressure gradients between the coal canister interior/exterior and significant concentration gradients within/outside the coal matrix. In contrast, sinusoidal injection's continuous gradual variation (0 → 200 ml/min) generates smaller-magnitude, shorter-duration gradients. Consequently, during high-flow phases (>100 ml/min), gas molecule (N_2 and CH_4) migration rates are relatively slower under sinusoidal injection compared to stepwise injection, while the opposite trend manifests during low-flow phases (<100 ml/min).

When injection flow rates vary cyclically between 175 and 25 ml/min and 150 and 50 ml/min, the outlet mixed gas flow rates and CH_4 flow rates under both variable-rate injection methods gradually converge over displacement time. This convergence arises from two primary factors: First, during low-to-high flow transitions in these ranges, stepwise injection maintains sustained flows at 175 or 150 ml/min. Though sinusoidal injection operates between 100 and 175 ml/min or 100 and 150 ml/min, the diminished flow differential reduces disparities in pressure gradients (across the coal canister) and concentration gradients (within the coal matrix) between the two methods. Second, during high-to-low flow transitions, stepwise injection maintains sufficient driving force at lower flow rates to sustain gas migration within the coal, preventing rapid attenuation of pressure and concentration gradients. Consequently, under cyclic injection conditions at 175–25 and 150–50 ml/min, the outlet mixed flow rates and CH_4 flow rates of stepwise and sinusoidal injection progressively converge over displacement time. During cyclic injection flow

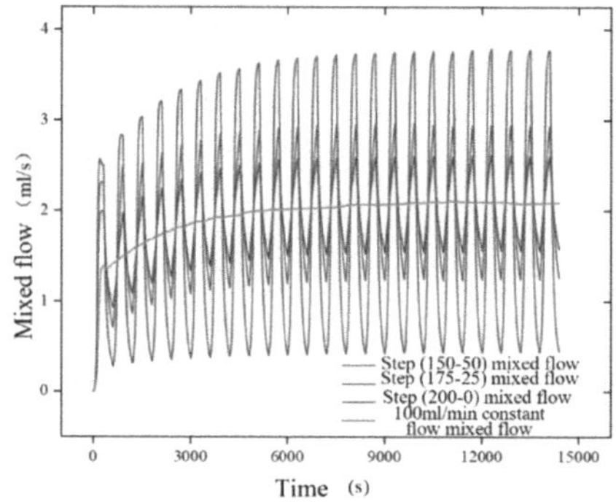

(a) Mixed Gas Outlet Flow Variation under Sinusoidal Injection

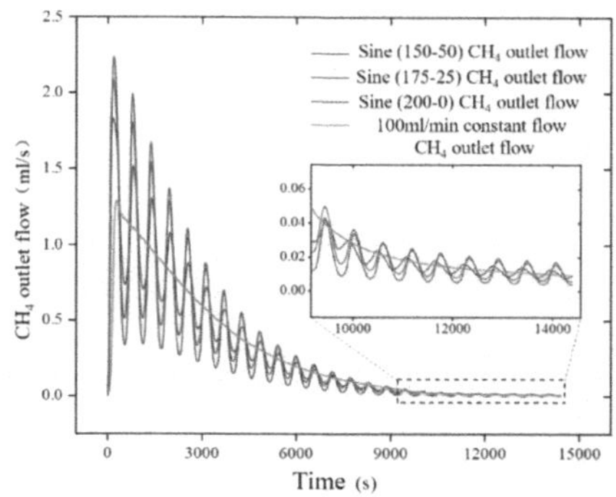

(b) CH_4 Outlet Flow Variation under Sinusoidal Injection

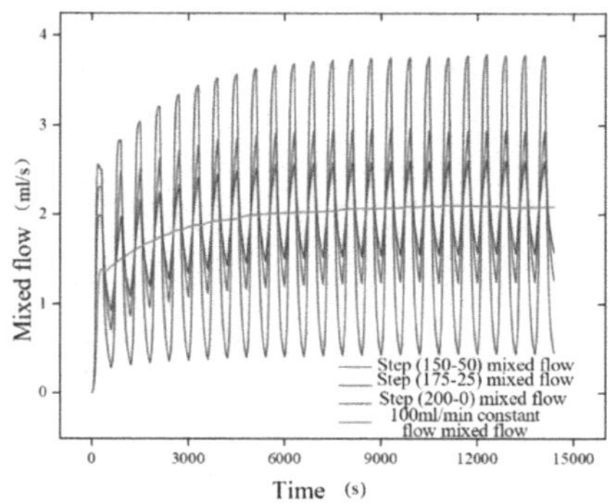

(c) Mixed Gas Outlet Flow Variation under Stepwise Injection

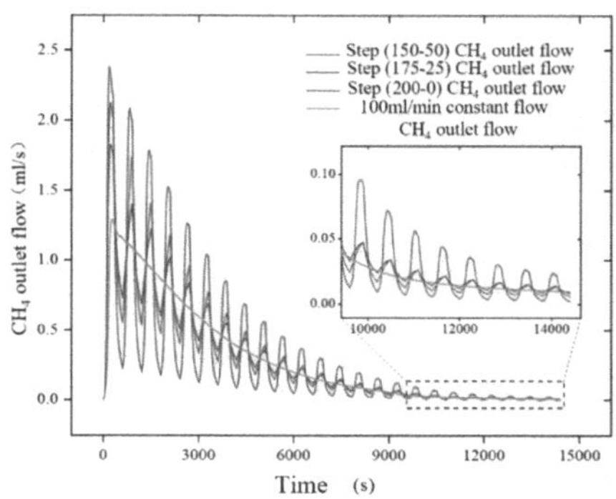

(d) CH_4 Outlet Flow Variation under Stepwise Injection

Fig. 4.20 Variation of mixing flow rate and CH_4 flow rate at the exit end of gas injection to replace coal grain CH_4 process under the conditions of different gas injection methods. (**a**) Mixed gas outlet flow variation under sinusoidal injection. (**b**) CH_4 outlet flow variation under sinusoidal injection. (**c**) Mixed gas outlet flow variation under stepwise injection. (**d**) CH_4 outlet flow variation under stepwise injection

rate variations within the 175–25 ml/min range, stepwise injection yields peak outlet mixed gas and CH_4 flow rates of 0.0495 and 0.0353 ml/s, respectively, with minimum values of 0.012 and 0.000108 ml/s; sinusoidal injection produces peak mixed gas and CH_4 flows of 0.0477 and 0.0328 ml/s, respectively, with minimum values of 0.010 and 0.00010 ml/s. During cyclic flow variations within the 150–50 ml/min range, stepwise injection achieves peak mixed gas and CH_4 flows of 0.0437 and 0.0305 ml/s, respectively, with minimum values of 0.0155 and 0.000122 ml/s; sinusoidal injection yields peak mixed gas and CH_4 flows of 0.0427 and 0.0305 ml/s, respectively, with minimum values of 0.0145 and 0.000118 ml/s. This data further demonstrates that convergence between stepwise and sinusoidal injection intensifies with narrower amplitude ranges—lower peak flow values combined with higher trough flow values progressively diminish differences in outlet mixed flow and CH_4 flow rates.

4.4.3.3 Variation in CH_4 Cumulative Emissions

Figure 4.21 illustrates the variation in CH_4 production under different gas injection methods. It can be observed that although cumulative CH_4 emissions vary across different injection methods and flow rates, the overall trend follows consistent patterns over the injection duration.

During Phase I (cycles 1–8), the cumulative CH_4 emissions exhibit a rapid increase. In Phase II (cycles 9–17), the cumulative emissions gradually level off. By Phase III (cycles 17–24), the cumulative emissions stabilize with minimal change. During Phase I, not only is free-state CH_4 flushed out from open spaces within the coal sample canister, but a portion of originally adsorbed CH_4 also desorbs into the free state and is subsequently expelled. In Phases II and III, the emitted CH_4 primarily consists of originally adsorbed CH_4 that has desorbed into the free state. By Phase III, the cumulative CH_4 emissions gradually level off and stabilize, remaining essentially constant thereafter.

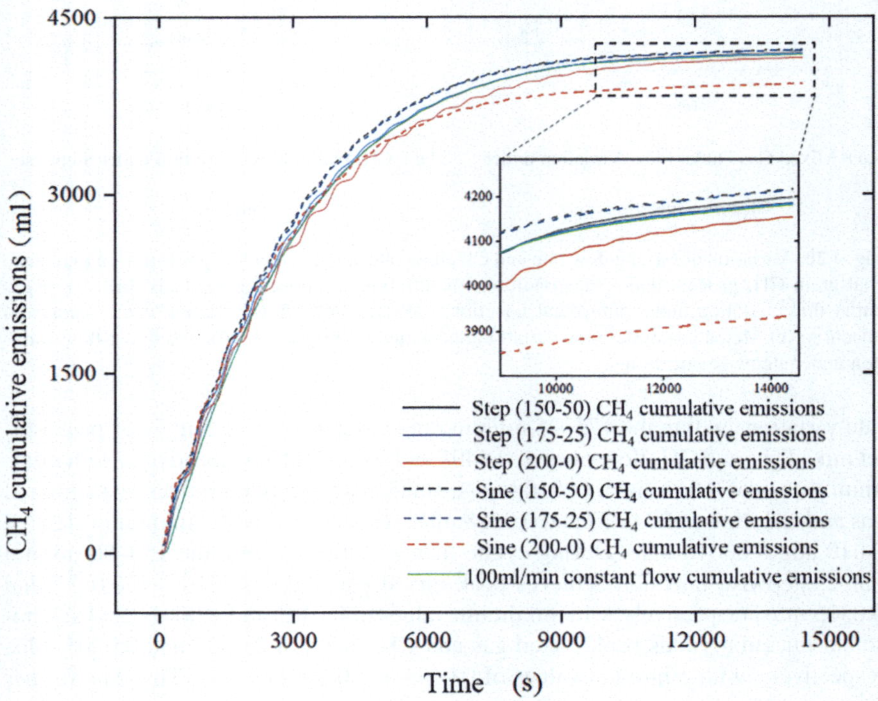

Fig. 4.21 Changes in cumulative CH_4 emissions by gas injection method

Figure 4.21 further demonstrates that under identical injection durations and flow conditions, the final cumulative CH_4 emissions exhibit distinct outcomes: under sinusoidal injection (200–0 ml/min), (175–25 ml/min), and (150–50 ml/min), the final cumulative CH_4 production volumes are 3933.30, 4213.69, and 4215.04 ml, respectively; while under stepwise injection (200–0 ml/min), (175–25 ml/min), and (150–50 ml/min), the corresponding values are 4154.62, 4186.08, and 4198.75 ml. Furthermore, under constant-flow injection at 100 ml/min, the final cumulative CH_4 production is 4183.01 ml. With identical injection duration but different injection methods, the CH_4 displacement efficiency ranks as follows from highest to lowest: sinusoidal (150–50 ml/min) > sinusoidal (175–25 ml/min) > stepwise (150–50 ml/min) > stepwise (175–25 ml/min) > 100 ml/min constant-flow > stepwise (200–0 ml/min) > sinusoidal (200–0 ml/min). This demonstrates that compared to constant-flow injection at 100 ml/min, sinusoidal (150–50 ml/min), sinusoidal (175–25 ml/min), stepwise (150–50 ml/min), and stepwise (175–25 ml/min) injection methods enhance CH_4 recovery from coal. The maximum cumulative CH_4 production across all methods is 1.0076 times that of the 100 ml/min constant-flow method. Experimental results demonstrate that when injection duration and average flow rate are identical, stepwise injection yields superior CH_4 recovery enhancement under 200–0 ml/min flow variations, whereas sinusoidal injection achieves better CH_4 recovery promotion under 175–25 ml/min and 150–50 ml/min flow ranges. Notably, for any given injection method, the 150–50 ml/min flow variation consistently produces higher final CH_4 emissions than other flow patterns. Hence, maintaining sustained injection gas flow during low-flow phases significantly enhances CH_4 recovery, with this effect being particularly pronounced under sinusoidal injection.

Therefore, under sinusoidal injection conditions, maintaining sustained N_2 injection during low-flow phases delivers more significant enhancement in coalbed methane recovery. Additionally, when injection method, duration, and average flow rate are identical, implementing 200–0 ml/min flow variations fails to increase CH_4 recovery while consuming equivalent N_2 volumes—resulting in resource waste that renders this approach unsuitable for N_2 injection in coalbed methane displacement. Furthermore, within sinusoidal injection protocols, preserving continuous N_2 injection during low-flow intervals proves particularly effective for boosting CH_4 production.

4.4.3.4 Influence of Injection Methods on CH_4 Emissions

As established previously, compared to constant-flow injection at 100 ml/min, sinusoidal (150–50 ml/min), sinusoidal (175–25 ml/min), stepwise (150–50 ml/min), and stepwise (175–25 ml/min) methods significantly enhance CH_4 recovery. Conversely, stepwise (200–0 ml/min) and sinusoidal (200–0 ml/min) methods demonstrate inferior displacement efficiency. However, when dividing the entire CH_4 emission process into discrete injection cycles (each cycle: 10 min duration, 1000 ml gas volume), distinct periodic emission patterns emerge across methods. For instance, Fig. 4.21 clearly shows that sinusoidal (200–0 ml/min) yields the lowest ultimate CH_4 recovery, yet achieves the highest cumulative emissions within the initial 2000 s. This indicates that different injection methods enhance CH_4 recovery during distinct temporal phases—a critical insight for optimizing gas source utilization and improving coalbed methane production efficiency, warranting further investigation.

A comparison of cumulative CH_4 emissions per cycle under various injection methods is presented in Fig. 4.22. It can be observed that except for constant-flow injection at 100 ml/min, all other injection methods exhibit decreasing CH_4 emissions with increasing cycle numbers. The CH_4 emissions during the second injection cycle of 100 ml/min constant-flow injection exceed those of the first cycle, primarily occurring within 600 s before and after N_2 breakthrough. Before breakthrough (within 300 s), continuous N_2 injection establishes gas concentration gradients within the coal, accelerating CH_4 migration and gradually increasing outlet CH_4 flow. After breakthrough (300–600 s), the total outlet CH_4 discharge exceeds the N_2 injection flow rate, causing delayed gas replenishment, reduced concentration gradients, diminished CH_4 desorption, and decreased outlet flow—consistent with earlier observations. Consequently, the initial phase of gradual CH_4 flow increase and the post-breakthrough decline predominantly occur during the first injection cycle. By the second cycle, this pattern disappears; outlet CH_4 emissions progressively increase and surpass the first cycle, resulting in higher emissions during the second injection cycle.

Figure 4.22 further reveals that during the first injection cycle, CH_4 emissions are lowest under constant-flow injection at 100 ml/min and highest under sinusoidal injection (150–50 ml/min). Additionally, sinusoidal injection generally yields higher CH_4 emissions. Since coal retains high initial CH_4 concentrations during the first cycle—with emissions comprising both free-state and partially adsorbed CH_4—sinusoidal injection proves more effective for recovering high-concentration CH_4. During cycles 2–9, the flushing efficiency of cyclic variable-rate injection gradually declines [3], whereas constant-flow injection at 100 ml/min achieves the highest CH_4 output. As emissions in this phase primarily originate from desorption of originally adsorbed CH_4, constant-flow injection demonstrates superior efficiency in liberating adsorbed CH_4 and enhancing recovery—consistent with prior research [3]. Contrasting literature [3], by cycle 10, CH_4 concentrations under most methods drop below 10%. Here, stepwise injection becomes significantly more effective: from cycles 10–21, it consistently outperforms other methods in CH_4 recovery. During cycles 22–24, although stepwise injection (175–25 ml/min) shows a slight decline, stepwise methods (200–0 and 150–50 ml/min) maintain higher emissions than alternatives. This underscores stepwise injection's particular suitability for enhancing recovery of difficult-to-desorb CH_4 during late-stage displacement.

The experimental results above indicate that with identical N_2 injection duration and average flow rate, appropriately controlling the injection method can significantly enhance CH_4 emission from coal and thereby improve recovery efficiency. For instance: in specific engineering contexts prioritizing high-concentration gas utilization—such as direct combustion for power generation or concentration enhancement for CNG/LNG production [198–204]—combining sinusoidal injection with constant-flow injection proves more effective for optimizing the efficiency and yield of high-concentration gas applications. Conversely, for objectives focused on mitigating coal seam gas hazards (e.g., preventing coal and gas outbursts), integrated injection strategies combining sinusoidal, constant-flow, and stepwise methods are expected to deliver superior outcomes. Additionally, recent advancements in low-concentration gas utilization (1–6%)—including technologies like regenerative thermal oxidation, porous media premixed combustion, and pulsating combustion [205]—reveal that stepwise injection also offers distinct advantages for enhancing both utilization and production efficiency of low-concentration gas under such conditions.

4.5 Macro-Microscopic Differences in CH_4 Dynamic Transport Behavior During Gas Injection and Displacement

Taking Sects. 2.5.3, 3.4.1, and 4.4.2 as examples, this study conducts a macro-microscopic differential analysis of CH_4 gas transport behavior during gas injection and displacement processes.

After expanding the research scale, certain discrepancies emerge between the variations in CH_4 adsorption capacity (Fig. 3.10) and pore-scale CH_4 molecular adsorption quantities (Fig. 2.21). Specifically, Fig. 2.21 demonstrates that during N_2 injection displacement, the adsorbed CH_4 molecules within pores exhibit an initial rapid decline followed by a slower decrease, whereas under CO_2 injection displacement, the reduction rate of adsorbed CH_4 molecules first proceeds more slowly before accelerating compared to N_2 injection. In contrast, Fig. 3.10 reveals consistently faster CH_4 desorption rates during CO_2 injection displacement relative to N_2 injection. This phenomenon arises because the current research scale extends beyond individual pores within the coal matrix to simultaneously consider desorption phenomena of adsorbed CH_4 gas occurring both inside and on the surface of the coal matrix. During gas injection displacement, adsorbed CH_4 molecules on the coal matrix surface desorb first. As evidenced by Figs. 3.8 and 3.9, injected CO_2 gas rapidly competes with adsorbed CH_4 molecules on the coal surface, resulting in greater desorption of surface-adsorbed CH_4 under CO_2 injection compared to N_2 injection, thereby explaining the observed desorption rate differences across scales. Notably, CO_2 injection displacement yields higher cumulative CH_4 desorption quantities at both scales compared to N_2 injection. In conclusion, while temporal variations in desorption rates of adsorbed CH_4 differ between scales during gas injection displacement, the total cumulative CH_4 desorption quantities remain consistent across measurement scales.

Furthermore, as established in previous sections, during gas injection and displacement processes at both macroscopic experimental and pore scales, free-state CH_4 is rapidly discharged within a short timeframe, while adsorbed-state CH_4 accounts for the majority of the emission duration. A comprehensive comparison of Figs. 2.22, 4.17, and 4.18 reveals that at both macroscopic and pore scales, the differences in CH_4 emission volumes and displacement durations remain consistent regardless of whether N_2 or CO_2 is injected. Given that the initial quantities of free-state CH_4 and adsorbed-state CH_4 (both within and on the coal matrix surface) are identical across all conditions, the variations in CH_4 transport behavior at the microscopic pore scale under different influencing factors can consequently manifest at the macroscopic scale. In other words, the impact of various factors on total CH_4 emission volumes and displacement durations at the macroscopic experimental scale is fundamentally determined by the corresponding CH_4 behavior at the microscopic pore scale under the influence of these same factors.

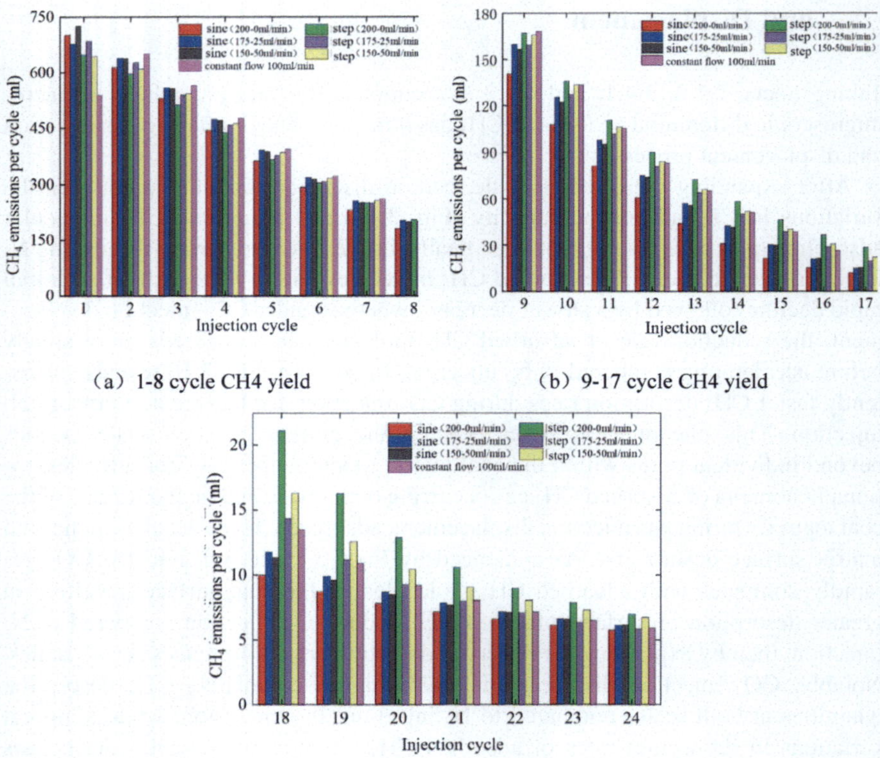

Fig. 4.22 Changes in CH_4 emissions in each cycle under the conditions of different gas injection methods. (**a**) 1–8 cycle CH_4 yield. (**b**) 9–17 cycle CH_4 yield. (**c**) 18–24 cycle CH4 yield

4.6 Characterization of CH_4 Dynamic Diffusion Behavior in Coal During Gas Injection and Displacement Using Diffusion Models

4.6.1 Classical Diffusion Models for Gas in Coal

The theoretical model of gas diffusion in coal serves as the foundation for studying gas diffusion behavior in coal formations. Currently established gas diffusion models in coal are primarily based on Fick's law of diffusion. Traditional diffusion models simplify the pore structure of coal into single-pore, dual-pore, or triple-pore systems. In this section, the widely applied single-pore and dual-pore diffusion models are selected to fit experimental data, with the aim of investigating the applicability of these gas diffusion models for describing CH_4 diffusion in coal during gas injection and displacement processes.

4.6.1.1 Single-Pore Diffusion Model

According to Fick's first law of diffusion, the mass flux through a given cross-section per unit time is proportional to the concentration gradient, as expressed in Eq. (4.2):

$$J = -D\frac{\partial C}{\partial x} \tag{4.2}$$

In the equation, J represents the mass of fluid passing through a unit cross-sectional area per unit time, in g/(m^2·s); D denotes the fluid diffusion coefficient, in m2/s; C is the fluid concentration, in mol/m^3; x is the fluid diffusion path, in m; $\partial C/\partial x$ indicates the concentration gradient; and the "—" sign signifies that the fluid diffusion direction is opposite to the concentration gradient direction.

Most practical diffusion processes involve non-steady-state diffusion, requiring consideration of temporal factors. Thus, Fick's first law of diffusion is refined into Fick's second law of diffusion. Assuming coal particles are isotropic spheres with radius r, the second law of Fickian diffusion for fluids in spherical coordinates is expressed by Eq. (4.3):

$$\frac{\partial C}{\partial t} = D\left(\frac{\partial^2 C}{\partial r^2} + \frac{2}{r}\frac{\partial C}{\partial r}\right) \tag{4.3}$$

Assuming the external gas concentration around the spherical coal particle remains constant and the internal gas concentration is C_1, the boundary conditions for Eq. (4.3) are given by Eqs. (4.4) and (4.5) as follows:

$$C = C_1, 0 < r\langle a, t \rangle 0 \tag{4.4}$$

$$C = C_0, r = a, t > 0 \tag{4.5}$$

In the equation, a denotes the radius of the spherical coal particle (unit: m), and C_0 represents the initial gas concentration in the particle (unit: mol/m^3).

Assuming instantaneous completion of gas adsorption and desorption processes with linear adsorption behavior in particles, the gas diffusion quantity depends solely on the concentration gradient. By employing the separation of variables method, the solution to Eq. (4.3) is obtained as Eq. (4.6), with detailed derivation available in Ref. [206].

$$\frac{C - C_1}{C_0 - C_1} = 1 + \frac{2a}{\pi r} \sum_{n=1}^{\infty} \frac{(-1)^n}{n} \sin \frac{n\pi r}{a} \times \exp\left(-Dn^2\pi^2 t / a^2\right) \tag{4.6}$$

In the equation, *n* represents the order of the infinite series terms.

The gas concentration at the center of the spherical particle can be obtained by taking the limit as $r \rightarrow 0$:

$$\frac{C - C_1}{C_0 - C_1} = 1 + 2\sum_{n=1}^{\infty} (-1)^n \times \exp\left(-Dn^2\pi^2 t / a^2\right) \tag{4.7}$$

Let m_t be defined as the total diffused mass of gas adsorbed/desorbed at time *t*, and m_∞ as the total diffused mass at complete adsorption/desorption equilibrium. The general expression for the diffusion process is then given by Eq. (4.8):

$$\frac{m_t}{m_\infty} = 1 - \frac{6}{\pi^2} \sum_{n=1}^{\infty} \frac{1}{n^2} \exp\left(-n^2\pi^2 t \frac{D}{r^2}\right) \tag{4.8}$$

Equation (4.8) represents the classical mathematical expression of the single-pore diffusion model.

In the equation, *t* represents the desorption/adsorption time (s), and the value of D/r^2 can be regarded as the effective diffusion coefficient D_e (1/s). In 1975, recognizing the limitations of the single-pore diffusion model assumptions, Crank et al. [206] proposed a modified calculation method for the single-pore diffusion model applicable to varying external concentrations. Although the heterogeneous nature of coal structure appears inherently incompatible with the single-pore diffusion model, experimental studies [207–209] have demonstrated that this model can still provide satisfactory fitting of diffusion data in certain cases.

4.6.1.2 Dual-Pore Diffusion Model

In contrast to the single-pore diffusion model which assumes a single-pore type in the coal matrix, the dual-pore diffusion model postulates the existence of two distinct pore structures—macropores and micropores—within the coal matrix, enabling more accurate characterization of gas diffusion phenomena in coal. The dual-pore diffusion model has been extensively applied in predicting coalbed methane production and evaluating coalbed methane resources, serving as a crucial tool for understanding and simulating coalbed methane extraction processes. This model categorizes the gas diffusion process in coal into macropore (rapid) diffusion and micropore (slow) diffusion components.

For the macropore diffusion stage, the diffusion model can be expressed as:

$$\frac{m_a}{m_{a\infty}} = 1 - \frac{6}{\pi^2}\sum_{n=1}^{\infty}\frac{1}{n^2}\exp\left(-n^2\pi^2 t\frac{D_a}{r_a^2}\right) \tag{4.9}$$

For the micropore diffusion stage, the diffusion model is expressed as:

$$\frac{m_i}{m_{i\infty}} = 1 - \frac{6}{\pi^2}\sum_{n=1}^{\infty}\frac{1}{n^2}\exp\left(-n^2\pi^2 t\frac{D_i}{r_i^2}\right) \tag{4.10}$$

In Eqs. (4.9) and (4.10) above:

D_a and D_i represent the diffusion coefficients for macropores and micropores, respectively; D_a/r_a^2 and D_i/r_i^2 may be regarded as the effective diffusion coefficients D_{ae} and D_{ie} (unit: s^{-1}).

Thus, the total adsorption/desorption capacity in coal is the sum of these two stages, which can be expressed by Eq. (4.11):

$$\frac{m_t}{m_\infty} = \frac{m_a + m_i}{m_{a\infty} + m_{i\infty}} = \beta\frac{m_a}{m_{a\infty}} + (1-\beta)\frac{m_i}{m_{i\infty}} \tag{4.11}$$

In the equation, $\beta = \frac{m_{a\infty}}{m_{a\infty} + m_{i\infty}}$ represents the percentage of gas adsorption/desorption quantity in macropores relative to the total adsorption/desorption quantity in all pores.

4.6.2 Applicability Analysis of Single-CH_4 Gas Diffusion Models in Coal

The calculation method from reference [194] was employed to determine the CH_4 desorption-diffusion quantity during the experiments, which was then used to further calculate the CH_4 diffusion ratio. In this section, the diffusion parameters of both models were adjusted to simulate the diffusion behavior of single-component CH_4 in coal. For both gas diffusion models mentioned above, setting the series summation term $n = 5$ ensures both the accuracy of model fitting results and reduced computational complexity. Therefore, n was set to 5 to obtain optimal diffusion parameters that best fit the experimental curves, thereby facilitating the applicability analysis of the diffusion models.

In this section, the applicability of both single-pore and dual-pore diffusion models for analyzing single-component CH_4 diffusion in coal was evaluated, as illustrated in Fig. 4.23. The results demonstrate that under various pressure levels, both diffusion models exhibit highly similar performance in their respective simulations. However, when compared to the single-pore model, the dual-pore model's simulation curve shows better agreement with the experimentally measured m_t/m_∞ curve. This improved accuracy stems from the dual-pore model's consideration of both macropores and micropores, which more realistically represents the coal sample's complex pore structure, as opposed to the single-pore model's idealized pore structure assumption. The corresponding macropore and micropore diffusion coefficients in the dual-pore model provide a more comprehensive explanation for the gas diffusion process in coal.

Figure 4.24 presents the variations in single-component CH_4 diffusion parameters in coal under different pre-adsorption gas pressures. For parameter D in the single-pore diffusion model, the CH_4 diffusion coefficients in coal were 4.2067e-10,

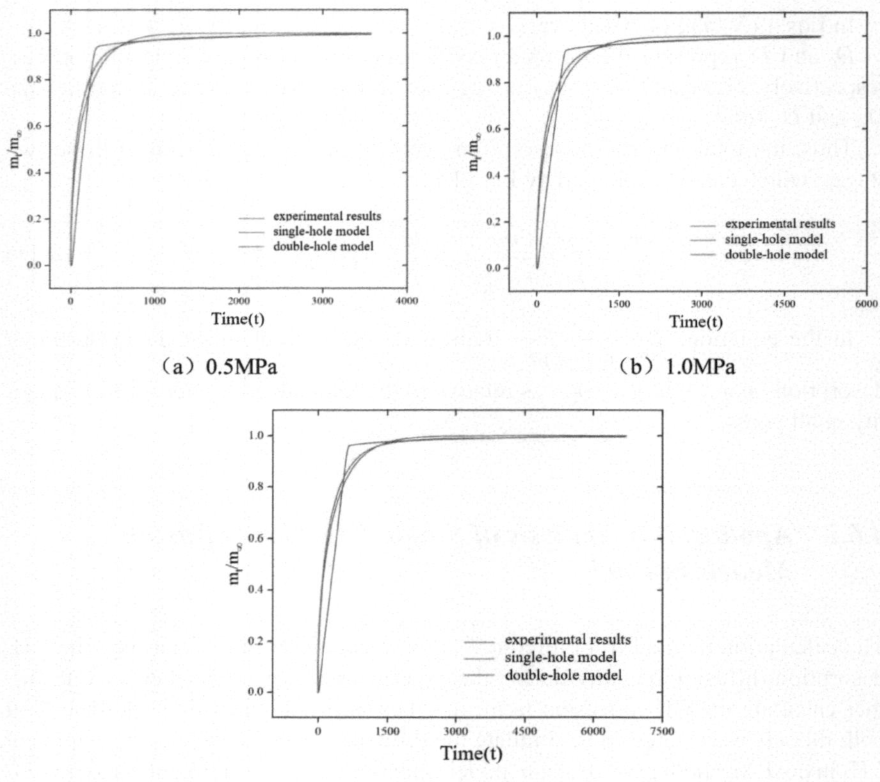

(a) 0.5MPa (b) 1.0MPa (c) 1.5MPa

Fig. 4.23 Fitting curve between experimental data and theoretical model of CH_4 diffusion. **(a)** 0.5 MPa. **(b)** 1.0 MPa. **(c)** 1.5 MPa

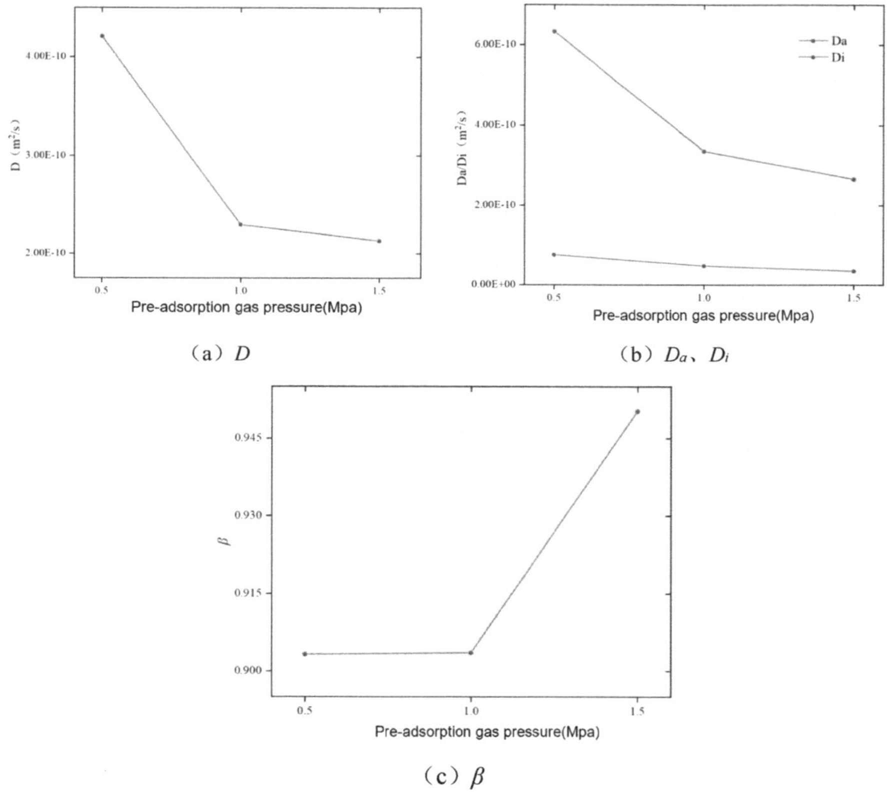

Fig. 4.24 Single CH_4 diffusion parameters for coal particles. (**a**) *D*, (**b**) D_a, D_i. (**c**) β

2.3017e-10, and 2.1275e-10 m^2/s at pre-adsorption gas pressures of 0.5, 1.0, and 1.5 MPa, respectively. Regarding parameters D_a and *Di* in the dual-pore diffusion model, the macropore diffusion coefficients of CH_4 in coal were 6.3345e-10, 3.3602e-10, and 2.6597e-10 m^2/s, while the micropore diffusion coefficients were 7.5834e-11, 4.8000e-11, and 3.5002e-11 m^2/s under the respective pressure conditions, all showing negative correlations with pre-adsorption gas pressure. The macropore proportion β of CH_4 diffusion in coal was 0.9033, 0.9036, and 0.9503 at 0.5, 1.0, and 1.5 MPa pre-adsorption pressures, respectively, exhibiting an increasing trend with rising pressure. These results demonstrate that for both single-pore and dual-pore diffusion models, the diffusion coefficients (D, D_a, and D_i) of single-component CH_4 in coal decrease with increasing pre-adsorption gas pressure, indicating that higher pressures lead to more complex diffusion pathways within the coal matrix and may even cause closure of some micropores. Furthermore, the reduction in diffusion coefficients was significantly greater when pressure increased from 0.5 to 1.0 MPa compared to the decrease observed from 1.0 to 1.5 MPa, suggesting that the influence of pre-adsorption gas pressure on single-component CH_4 diffusion coefficients has certain limitations.

4.6.3 Applicability Analysis of CH_4 Diffusion Models in Gas Injection and Displacement

4.6.3.1 Diffusion Kinetic Curves of CH_4 in Gas Injection and Displacement

During the gas injection and displacement experiments, the calculation method for CH_4 diffusion ratio (m_t/m_∞) remained consistent with that described in Sect. 4.6.2. Figure 4.25 presents the variation patterns of CH_4 diffusion ratio over time under different gas injection flow rates. The results demonstrate that gas diffusion ratios under various injection flow rates follow similar trends, with steep initial curve slopes that gradually flatten as injection time progresses. This indicates that the CH_4 diffusion ratio exhibits a rapid-slow-stable transition pattern with increasing injection time. Higher injection flow rates correspond to faster attainment of CH_4 diffusion ratio reaching unity, suggesting that increased injection flow rates accelerate CH_4 diffusion in coal. Furthermore, compared to N_2 injection displacement, when CO_2 is selected as the injection gas source under identical flow rate conditions, the CH_4 diffusion ratio in coal reaches unity earlier, demonstrating that CO_2 enhances the CH_4 diffusion rate in coal more effectively than N_2.

4.6.3.2 Applicability Analysis of CH_4 Diffusion Models in Gas Injection and Displacement

Following the same methodology as described in Sect. 4.6.2, the applicability of CH_4 diffusion models during gas injection and displacement was analyzed. The fitted curves are presented in Fig. 4.26.

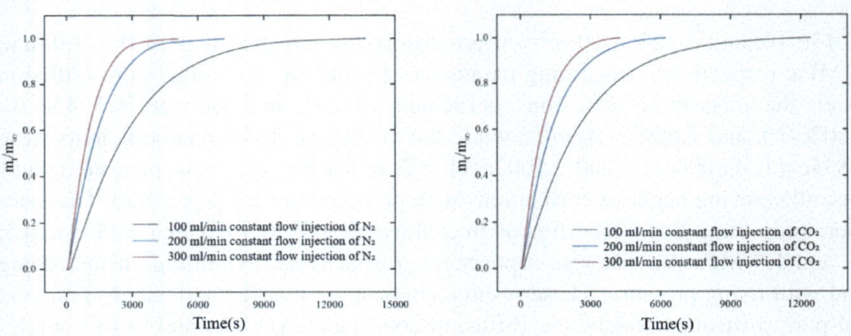

(a) The diffusion kinetic curves of CH_4 during N_2 injection displacement in coal

(b) The diffusion kinetic curves of CH_4 during CO_2 injection displacement in coal

Fig. 4.25 Fitting of CH_4 diffusion kinetic curves for gas injection-driven coal particles. (**a**) The diffusion kinetic curves of CH_4 during N_2 injection displacement in coal. (**b**) The diffusion kinetic curves of CH_4 during CO_2 injection displacement in coal

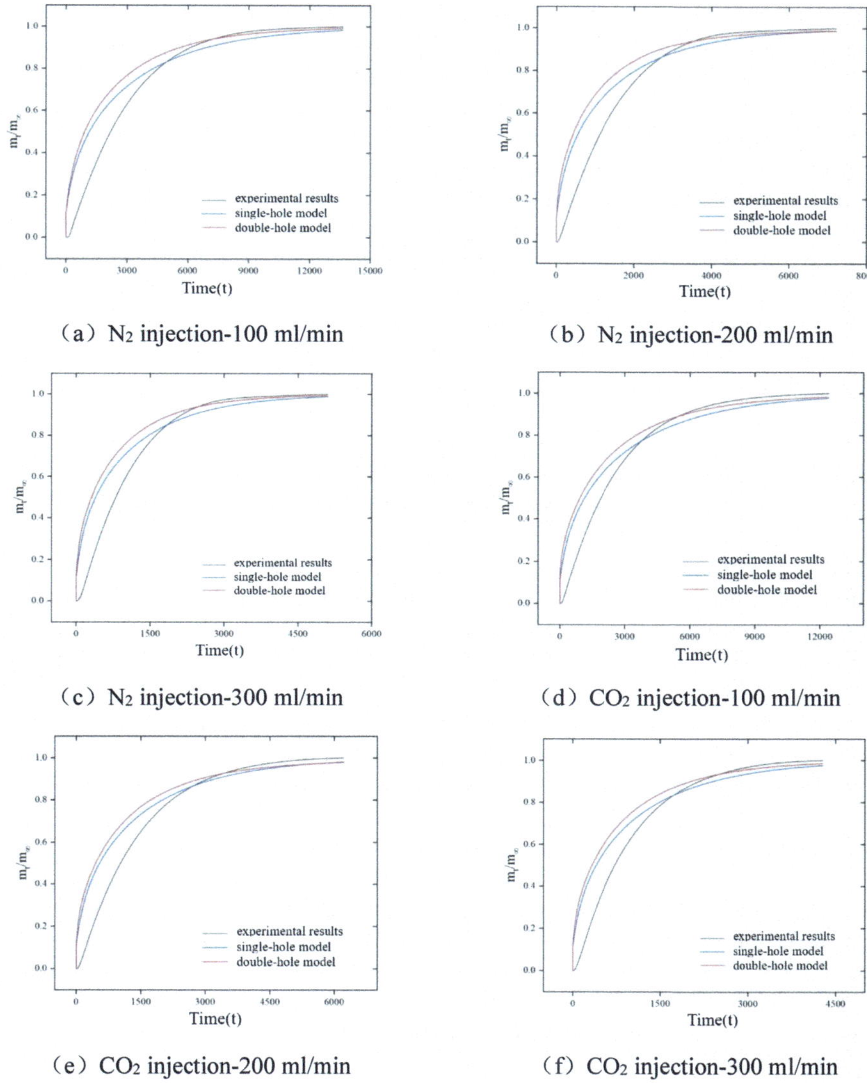

Fig. 4.26 Fitting of CH_4 diffusion kinetic curves for gas injection-driven coal particles. (**a**) N_2 injection—100 ml/min. (**b**) N_2 injection—200 ml/min. (**c**) N_2 injection—300 ml/min. (**d**) CO_2 injection—100 ml/min. (**e**) CO_2 injection—200 ml/min. (**f**) CO_2 injection—300 ml/min

It can be observed that for both N_2 and CO_2 injection displacement processes, when comparing the experimentally measured m_t/m_∞ curves, the single-pore diffusion model shows better agreement with the experimental curves in the early stage, while the dual-pore diffusion model demonstrates closer alignment in the later stage. This phenomenon is related to the diffusion pathways of CH_4 in coal: during the initial diffusion stage, CH_4 primarily desorbs from the surface or near-surface regions of coal, where the diffusion pathways are relatively short and the pore structure is simpler and more homogeneous. Consequently, the single-pore diffusion model effectively describes this straightforward diffusion process, resulting in closer agreement between the model fitting curve and the experimental data. As time progresses, the adsorbed CH_4 in the surface and near-surface regions gradually depletes, and the desorption-diffusion behavior of CH_4 extends from the surface to the deeper pore structures within the coal matrix. Simultaneously, CH_4 must undergo slow desorption-diffusion from micropores, leading to more complex diffusion pathways and reduced diffusion rates. Compared to the single-pore diffusion model, the dual-pore diffusion model provides a more accurate description of this complex diffusion process, particularly when accounting for the distinct diffusion characteristics of macropores and micropores, thereby achieving better fitting with the experimental data.

4.6.3.3 Comparative Analysis of Dynamic CH_4 Diffusion Parameters During Displacement and Differences from Pure CH_4 Diffusion

Figure 4.27 presents the variations of CH_4 diffusion parameters (D, D_a, D_i, β) with different gas injection flow rates during the displacement process. For the parameter D in the single-pore diffusion model, when N_2 is used as the injection gas, the CH_4 diffusion coefficients in coal at flow rates of 100, 200, and 300 ml/min are 2.7216e-11, 5.6986e-11, and 7.8166e-11 m^2/s, respectively. When CO_2 is used as the injection gas, the corresponding CH_4 diffusion coefficients are 2.6569e-11, 5.6645e-11, and 7.7607e-11 m^2/s, slightly lower than those observed with N_2 injection. Furthermore, the CH_4 diffusion coefficients increase progressively with higher injection flow rates.

For the dual-pore diffusion model, under N_2 injection displacement at flow rates of 100, 200, and 300 ml/min, the macropore diffusion coefficients of CH_4 in coal were 3.5123e-11, 7.6768e-11, and 9.9981e-11 m^2/s, respectively, while the micropore diffusion coefficients were 2.0176e-11, 2.2029e-11, and 4.5166e-11 m^2/s, respectively. The macropore proportion (β) of CH_4 diffusion in coal was 0.9002, 0.9328, and 0.9511, respectively. Under CO_2 injection displacement at the same flow rates, the macropore diffusion coefficients of CH_4 in coal were 3.4755e-11, 7.4381e-11, and 9.5236e-11 m^2/s, respectively, while the micropore diffusion coefficients were 1.6342e-11, 2.1035e-11, and 4.3019e-11 m^2/s, respectively. The macropore proportion (β) was 0.8931, 0.9117, and 0.9319, respectively—all lower than those under N_2 injection displacement in terms of both macropore and micropore diffusion coefficients and macropore proportion. This is because CO_2, as a strongly adsorbable gas, induces adsorption-induced swelling of the coal matrix when injected, thereby reducing the diffusion coefficients of CH_4. This phenomenon is demonstrated in Fig. 4.23c, where the macropore proportion under CO_2 injection displacement is lower than that under N_2 injection displacement. This indicates that CO_2-induced adsorption swelling reduces the pore size of the coal matrix, decreasing the proportion of macropores. Additionally, as the injection flow rate increases,

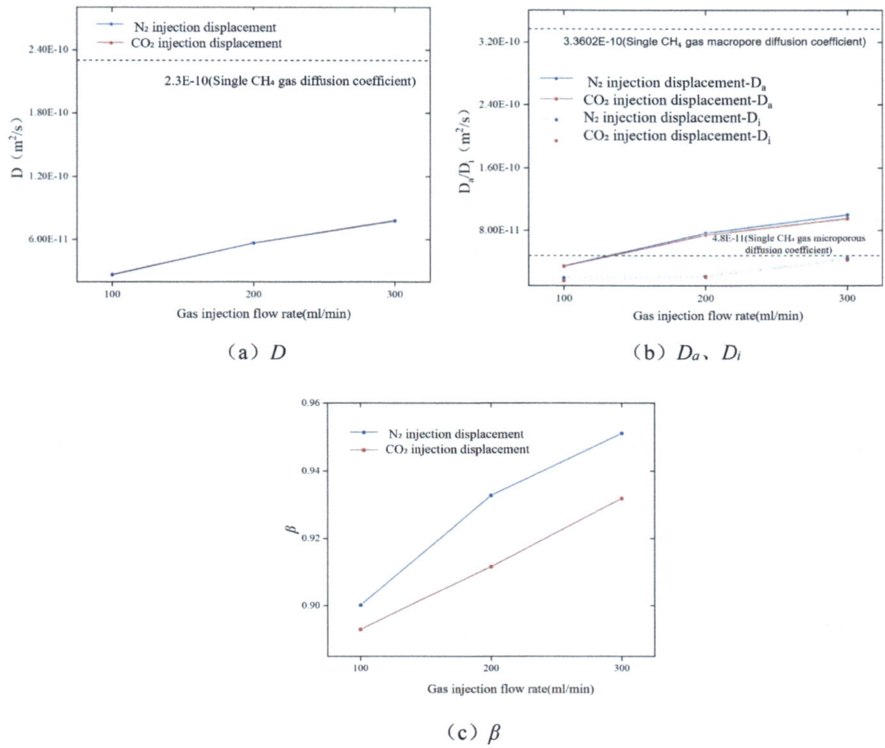

Fig. 4.27 CH_4 diffusion parameters of gas injection-driven coal particles. (**a**) D. (**b**) D_a, D_i. (**c**) β

both the macropore and micropore diffusion coefficients of CH_4, as well as the macropore proportion, exhibit an increasing trend.

However, both the single-pore and dual-pore diffusion models yield lower diffusion coefficients for CH_4 compared to those under single-component CH_4 diffusion at equivalent pre-adsorption gas pressures. This indicates that the injection of foreign gases during gas injection and displacement reduces the diffusion coefficient of CH_4, as the "diffusion-adsorption" behavior of the injected foreign gases in coal partially inhibits the "desorption-diffusion" rate of CH_4.

Furthermore, in comparison to the dynamic diffusion of CH_4 during the displacement process, the normalized uptake ratio (m_t/m_∞) of single-component CH_4 diffusion increases more rapidly in the early stage but slows down in the later stage. This occurs because, in single-component CH_4 diffusion, the coal sample is directly exposed to atmospheric conditions, resulting in a significant initial pressure gradient between the interior and exterior of the sample chamber. This leads to rapid gas flow and a sharp initial concentration gradient of adsorbed CH_4 in the coal matrix, accelerating the desorption-diffusion rate in the early stage. However, as diffusion progresses, the pressure gradient quickly diminishes to zero, halting CH_4 flow and causing the adsorbed CH_4 concentration gradient to also drop rapidly to zero, significantly slowing the desorption-diffusion rate in the later stage. In contrast, during the displacement process, the continuous injection of foreign gases maintains a sustained concentration gradient for adsorbed CH_4 both inside and outside the coal matrix, thereby driving the desorption-diffusion of adsorbed CH_4.

4.6.4 Introduction and Applicability Analysis of the Diffusion-Relaxation Model

4.6.4.1 Introduction of the Diffusion-Relaxation Model

Limitations of Classical Gas Diffusion Models

Previous studies have shown that single-pore and dual-pore gas diffusion models can effectively fit experimental data for single-component gases [210–212]. However, as discussed in Sects. 4.6.2 and 4.6.3, both the single-pore and dual-pore diffusion models exhibit significant deviations when fitting CH_4 diffusion experimental data under gas injection and displacement conditions. Specifically, in the early stages, the calculated results from these models exceed the experimental measurements, whereas in the later stages, the model predictions fall below the experimental values. This discrepancy indicates that traditional single-pore and dual-pore gas diffusion models are inadequate for describing CH_4 diffusion behavior under gas injection and displacement conditions. The primary reasons include the following:

1. Single-pore and dual-pore diffusion models primarily consider the diffusion behavior of a single gas, making them inherently unsuitable for describing CH_4 diffusion in coal under conditions involving multicomponent gas interactions during gas injection and displacement.
2. Classical gas diffusion models assume homogeneity in both coal and pore systems, which contradicts experimental observations, as demonstrated in prior research [55].
3. Previous studies have confirmed that gas adsorption induces coal matrix swelling, leading to non-Fickian anomalous diffusion behavior [21, 147, 213], a factor not accounted for in single-pore or dual-pore models.
4. During gas injection and displacement, the increasing presence of foreign gases in deeper coal matrix pores complicates the diffusion pathways of CH_4, resulting in continuously varying diffusion coefficients that cannot be represented by one or a few constant values.

Construction of the Diffusion-Relaxation Model

The establishment of a theoretical model must be based on reasonable assumptions. The initial assumptions of the diffusion-relaxation model include the following aspects: (1) The time required for gas adsorption/desorption in the coal matrix is negligible, and the desorbed free gas follows the ideal gas equation of state; (2) The transport of gas in the coal matrix adheres to the mass conservation and n of gas in the coal matrix; (3) The external gas concentration of the coal matrix remains constant.

The diffusion-relaxation model combines Fickian diffusion with a first-order relaxation model, initially proposed by Berens and Hopfenber [214] to describe the rapid diffusion and slow relaxation processes in glassy polymer systems. This model posits that the diffusion process of adsorbates in the adsorbent is a linear superposition of two independent phenomena: Fickian rapid diffusion and slow relaxation, as expressed in Eq. (4.12):

$$m_t = m_{t,F} + m_{t,R} \tag{4.12}$$

In the equation, m_t represents the total diffusion amount of CH_4 gas in the coal matrix at time t; $m_{t,F}$ denotes the Fick diffusion component of CH_4 gas in the coal matrix at time t; and $m_{t,R}$ indicates the slow relaxation component of CH_4 gas diffusion in the coal matrix at time t.

For the Fick diffusion component, it is expressed using the single-pore diffusion model, as shown in Eq. (4.13):

$$m_{t,F} = m_{\infty,F}\left[1-\frac{6}{\pi^2}\sum_{n=1}^{\infty}\frac{1}{n^2}\exp\left(-n^2\pi^2 t\frac{D}{r^2}\right)\right] \tag{4.13}$$

In the equation, $m_{\infty,F}$ represents the limiting diffusion amount in the easily accessible diffusion regions.

For the slow relaxation component, assuming the relaxation process follows a first-order kinetics driven by the concentration difference, the differential equation governing the relaxation process is given by Eq. (4.14):

$$\frac{dm_{t,R}}{dt} = k\left(m_{\infty,R} - m_{t,R}\right) \tag{4.14}$$

In the equation, k is the relaxation rate constant, and $m_{\infty,R}$ denotes the limiting diffusion amount induced by the relaxation phenomenon.

Integrating Eq. (4.14) yields Eq. (4.15):

$$m_{t,R} = m_{\infty,R}\left[1-\exp(kt)\right] \tag{4.15}$$

Substitution of Eqs. (4.13) and (4.15) into Eq. (4.12) yields the diffusion rate of the adsorbate at time t:

$$\frac{m_t}{m_\infty} = \varphi_F\left[1-\frac{6}{\pi^2}\sum_{n=1}^{\infty}\frac{1}{n^2}\exp\left(-n^2\pi^2\frac{D}{r^2}t\right)\right] + \varphi_R\left[1-\exp(-kt)\right] \tag{4.16}$$

$$\phi_F = \frac{m_{\infty,F}}{m_\infty} \tag{4.17}$$

$$\phi_R = \frac{m_{\infty,R}}{m_\infty} \tag{4.18}$$

In the equation, φ_F represents the diffusion contribution from Fick diffusion; φ_R denotes the diffusion contribution from relaxation diffusion, $\varphi_R = 1 - \varphi_F$; k is the first-order relaxation coefficient; and D/r^2 is defined as the effective diffusion coefficient D_e (unit: 1/s). When $D_e \ll k$, it represents pure relaxation diffusion; when $D_e \gg k$, it indicates pure Fick diffusion; intermediate values correspond to a third diffusion type termed anomalous diffusion.

4.6.4.2 Applicability Analysis of Diffusion-Relaxation Model

This section examines the applicability of the diffusion-relaxation model for characterizing CH_4 dynamic diffusion during CO_2 injection displacement in coal. The analytical solution of Eq. (4.16) was fitted to experimental measurements, with results presented in Fig. 4.28. Compared to single-porosity and dual-porosity diffusion models, the diffusion-relaxation model demonstrates superior fitting performance for experimental CH_4 diffusion data during CO_2 injection, achieving consistently high R^2 values of 0.999 (Table 4.7), confirming its capability to accurately represent CH_4 dynamic diffusion behavior under CO_2 injection conditions.

Table 4.7 further reveals that the relaxation coefficient k exceeds the effective diffusion coefficient D_e by 1–2 orders of magnitude across all tested injection flow rates. This indicates that relaxation diffusion dominates CH_4 dynamic diffusion during CO_2 injection displacement, primarily due to adsorption-induced matrix swelling during CO_2–CH_4 exchange. The CO_2 displacement process involves simultaneous diffusion of CO_2 from fractures into the coal matrix and desorption/diffusion of CH_4 from the matrix to fractures, coupled with competitive adsorption of both gases within the matrix pores. CO_2 adsorption induces significant matrix swelling that reduces pore size and narrows pore throats, increasing the energy barrier for molecular diffusion through micropores and consequently decreasing diffusion coefficients. This phenomenon prolongs the residence time of adsorbed gas in micropores, creating a gradual release process for CH_4 molecules trapped in swollen matrix pores.

Concurrently, Fickian diffusion driven by CH_4 concentration gradients between matrix surfaces and fractures must also be considered. Therefore, CH_4 diffusion during CO_2 injection comprises both concentration-driven Fickian diffusion and non-Fickian anomalous diffusion induced by adsorption swelling. This latter component, arising from matrix volume changes, is specifically defined as relaxation diffusion. In essence, CH_4 dynamic diffusion under CO_2 injection represents a superposition of Fickian and relaxation diffusion processes, consistent with the fundamental principles of the diffusion-relaxation model.

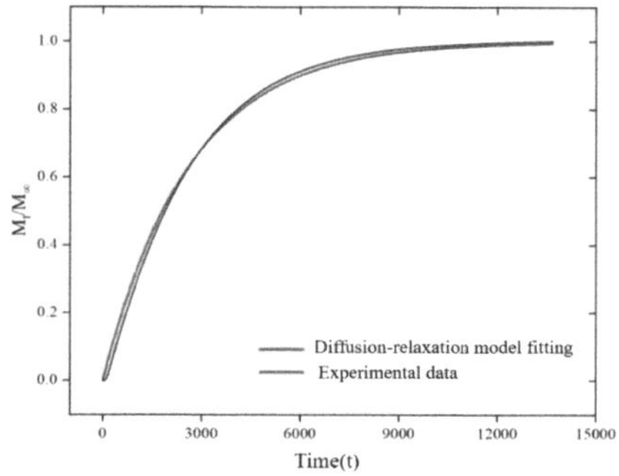

(a) 100 ml/min

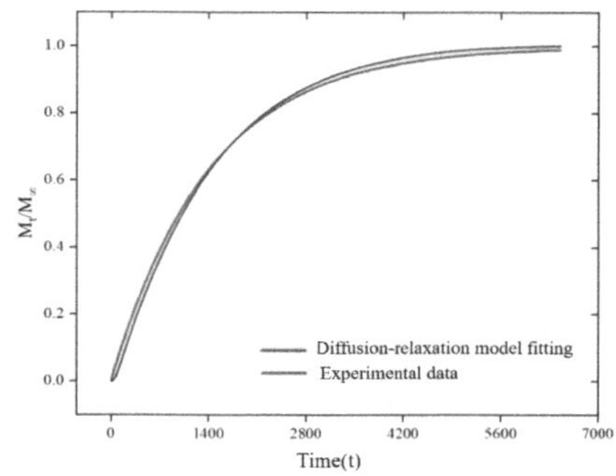

(b) 200 ml/min

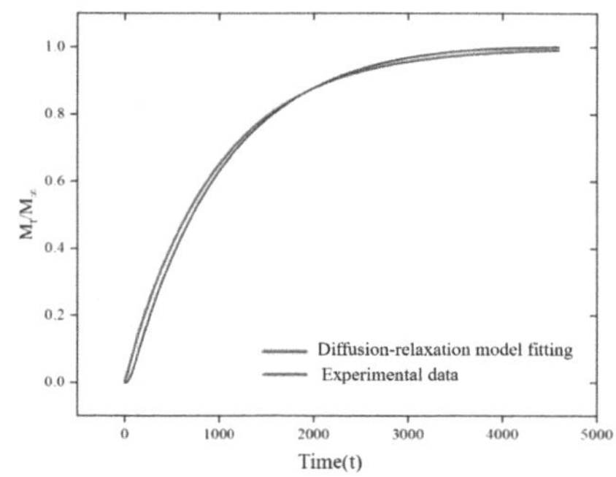

(c) 300 ml/min

Fig. 4.28 Diffusion-relaxation model fitting results to experimental data. (**a**) 100 ml/min. (**b**) 200 ml/min. (**c**) 300 ml/min

Table 4.7 Data fitting results

	Parameter				
Gas injection flow rate(ml/min)	φ_F	φ_R	D_e	k	R^2
100	0.0005	0.9995	4.14e-05	3.83e-04	0.999
200	0.0005	0.9995	5.08e-05	7.14e-04	0.999
300	0.0005	0.9995	6.48e-05	1.01e-03	0.999

4.6.4.3 Effect of Gas Injection Flow Rate on D_e and k

As previously described, increasing the injection flow rate also leads to higher inlet and outlet pressures. There exists controversy in existing literature regarding the relationship between single-gas diffusion coefficients and pressure. For instance, Staib et al. [215] suggest that both macropore and micropore diffusion coefficients of gases in coal decrease with increasing pressure, while Li et al. [216] propose that the effective diffusion coefficient increases with pressure. However, these findings only describe the variation patterns of single-gas diffusion coefficients and cannot be directly applied to characterize CH_4 diffusion coefficient changes under gas injection displacement conditions. Gang et al. [196] indicate that increasing injection pressure before CO_2 breakthrough can enhance the effective diffusion coefficient of CH_4, but finds no significant correlation between CH_4 diffusion coefficient and injection pressure post-breakthrough. Nevertheless, this study merely derived temporal variations of effective diffusion coefficients through mathematical transformations of the single-porosity diffusion model, which fails to represent true diffusion behavior. Consequently, it cannot accurately describe the CH_4 diffusion process under gas injection displacement conditions either.

The experimental data and analytical solution fitting results are presented in Fig. 4.29. Figure 4.29a demonstrates that the effective diffusion coefficient D_e increases with higher injection flow rates. At an injection flow rate of 300 ml/min, the maximum effective diffusion coefficient reaches 6.5e-5, which is 1.57 times greater than that at 100 ml/min. This phenomenon primarily relates to the molecular mean free path [217] (Fig. 4.30), where a larger λ corresponds to greater diffusion resistance and slower diffusion velocity for CH_4 in coal. As shown in Fig. 4.30, with increasing injection duration, higher injection flow rates result in smaller mean free paths for CH_4 molecules, thereby facilitating CH_4 diffusion. Consequently, greater injection flow rates lead to higher CH_4 diffusion coefficients.

Figure 4.29b illustrates the relationship between the relaxation coefficient k and the gas injection flow rate. The results demonstrate that higher injection flow rates correspond to larger k values. This phenomenon primarily occurs because, as evident from Eq. (4.16), a larger k value indicates a greater contribution from relaxation diffusion, reflecting more pronounced non-Fickian anomalous behavior induced by adsorption-induced coal matrix deformation. Given that coal exhibits stronger adsorption affinity for CO_2 than CH_4, higher CO_2 injection flow rates introduce more CO_2 molecules into the coal matrix per unit time, resulting in greater swelling deformation due to CO_2 adsorption. Consequently, the resulting relaxation effects from this swelling deformation become more significant, leading to higher values of the relaxation coefficient k.

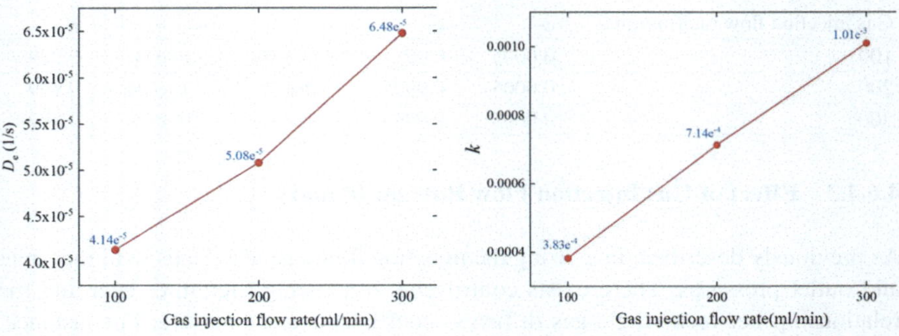

(a) Change rule of D_e (b) Change rule of k

Fig. 4.29 Change rule of D_e and k under different gas injection flow conditions. (**a**) Change rule of D_e. (**b**) Change rule of k

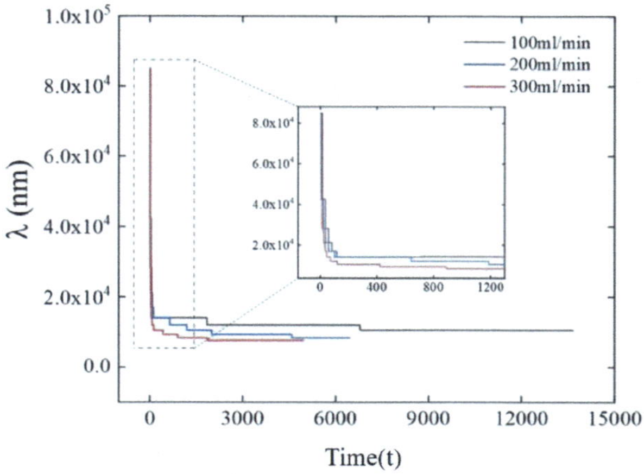

Fig. 4.30 Law of change of λ with time

4.6.5 *Dynamic Diffusion Mechanism of CH_4 During Gas Injection and Displacement Based on Diffusion-Relaxation Characteristics*

This section analyzes the dynamic diffusion mechanism of CH_4 during CO_2 injection displacement based on diffusion-relaxation characteristics, using 200 ml/min CO_2 injection for CH_4 displacement in coal as an example. As shown in Fig. 4.31, consistent with previous descriptions, both single-porosity and dual-porosity gas diffusion models demonstrate similar fitting performance for experimental results of CH_4 dynamic diffusion during CO_2 injection displacement. Specifically, during the early stages of gas injection displacement (encompassing both Phase I and Phase II), the simulation results of both diffusion models closely match experimental data in Phase I but significantly overestimate the results in Phase II. Conversely, during the later stage of gas injection displacement (Phase III), the simulation results from both single-porosity and dual-porosity diffusion models become notably lower than the experimental measurements.

During Phase I, the injected CO_2 increases gas pressure within the coal mass, creating a pressure gradient with the external environment that serves as the driving force to expel free-phase CH_4 from coal fractures. This process establishes an elevated concentration gradient of CH_4 and CO_2 between the coal matrix surface and free space, triggering rapid desorption of adsorbed CH_4 and simultaneous CO_2 adsorption at the matrix surface. The primary diffusion domain during this phase occurs at the coal matrix surface, where concentration differences between the matrix surface and fractures drive CH_4/CO_2 movement along their respective concentration gradients, conforming to Fick's diffusion law.

In Phase II, as displacement progresses, substantial CO_2 molecules penetrate into the microporous structure of the coal matrix interior. The accumulating CO_2

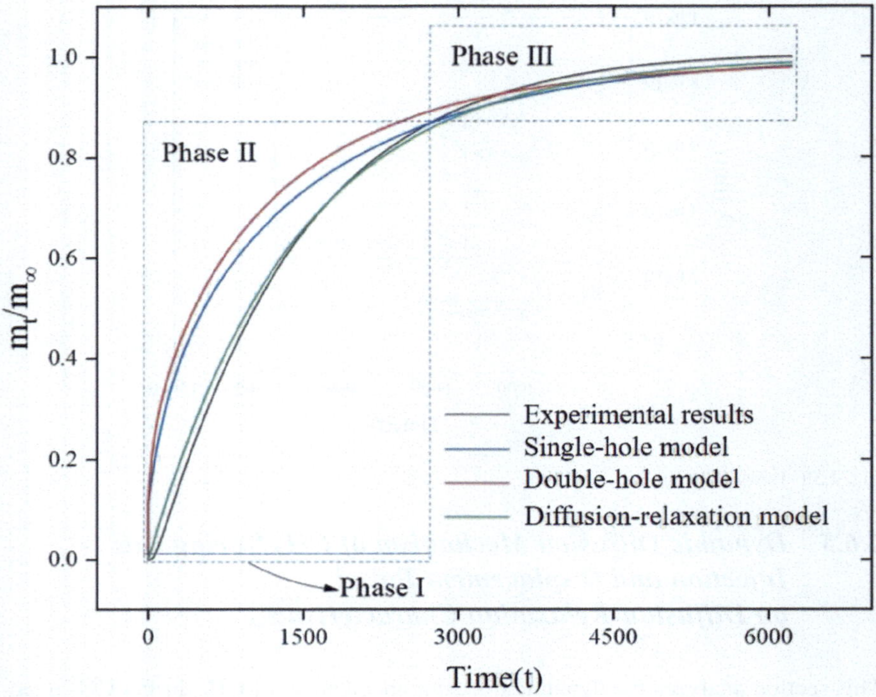

Fig. 4.31 Results of fitting the three diffusion models to the experimental data

adsorption induces matrix swelling deformation, thereby increasing diffusion resistance for CH_4. Previous studies have demonstrated that matrix adsorption swelling reduces gas diffusion rates [53]. However, both single-porosity and dual-porosity diffusion models neglect this adsorption-induced swelling effect—the diffusion relaxation phenomenon. Consequently, during Phase II, these classical diffusion models inevitably overestimate CH_4 dynamic diffusion behavior compared to experimental results, indicating that traditional gas diffusion models overrate CH_4 diffusion capacity in coal during early displacement stages.

During Phase III, the desorption of adsorbed CH_4 progressively extends into the interior of the coal matrix. As shown in Fig. 4.32, the CH_4 diffusion kinetic curve in Phase III gradually flattens, with its slope absolute value being smaller than that in Phase II. This phenomenon primarily occurs because prolonged CO_2 injection reduces the CH_4 concentration gradient between the coal matrix interior and fractures as CH_4 continuously desorbs and diffuses, thereby decreasing the driving force for CH_4 desorption. The matrix swelling induced by CO_2 adsorption exceeds the matrix shrinkage caused by CH_4 desorption [178, 218, 219], resulting in narrowed pores within the coal matrix and more complex gas diffusion pathways. This adsorption-induced swelling leaves additional CH_4 gas trapped in the microporous structure of the coal matrix that requires gradual desorption, causing the

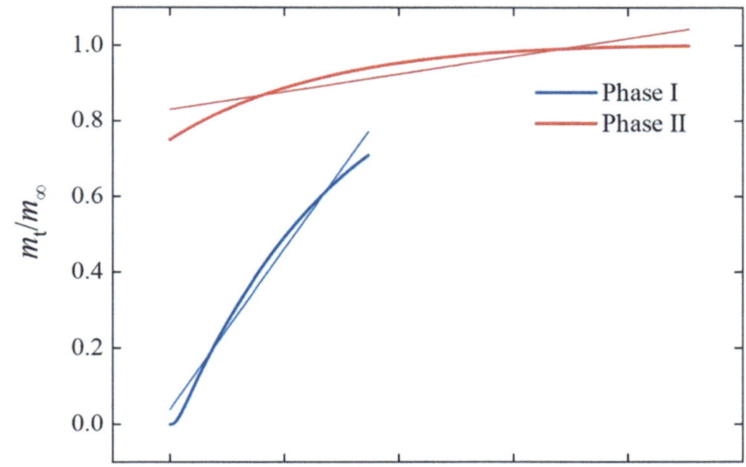

Fig. 4.32 Phase II and Phase III diffusivity changes

single-porosity and dual-porosity diffusion models to underestimate the experimental results when simulating CH_4 dynamic diffusion during Phase III. Consequently, applying these models to analyze CH_4 dynamic diffusion behavior during this stage would underestimate CH_4's diffusion capacity in coal.

In summary, since CO_2 injection displacement induces coal matrix adsorption swelling leading to anomalous diffusion phenomena, the conventional single-porosity and dual-porosity diffusion models based on Fick's law are no longer suitable for comprehensively describing CH_4 dynamic diffusion behavior under CO_2 injection conditions. It can be concluded that the CH_4 dynamic diffusion process during CO_2 injection displacement can be divided into two distinct components according to the locations where desorption-diffusion occurs: the first component involves Fickian diffusion of adsorbed CH_4 on coal matrix surfaces driven by concentration gradients during early displacement stages; the second component consists of relaxation diffusion within the coal matrix interior caused by adsorption swelling phenomena during middle and late displacement stages. The complete CH_4 dynamic diffusion process under CO_2 injection displacement conditions represents the superposition of these two components.

4.7 Chapter Summary

This chapter employs a self-developed experimental system for measuring dynamic transport characteristics of CH_4 during gas injection displacement to quantitatively investigate the dynamic transport patterns of CH_4 under various macroscopic conditions. Building upon macroscopic experimental results, the study theoretically examines the applicability of different single-gas diffusion models in characterizing

CH_4 dynamic diffusion behavior during displacement processes. The research systematically evaluates the influence of injection flow rates and gas source types on diffusion parameter variations across these models while investigating the dynamic diffusion mechanisms of CH_4 during gas injection displacement based on diffusion-relaxation characteristics. Furthermore, comparative analysis of CH_4 diffusion behavior under pure CH_4 diffusion versus displacement conditions was conducted based on experimental results, with the following specific conclusions obtained:

1. During the gas injection displacement process for enhancing CH_4 recovery from coal, increasing the injection flow rate leads to earlier breakthrough of foreign gas and a more rapid increase in the pressure gradient between the interior and exterior of the coal sample container. Concurrently, both the mixed gas flow rate and CH_4 outflow rate at the outlet exhibit faster increases and higher peak values. Throughout the displacement process, CH_4 gas undergoes distinct phases: breakthrough followed by rapid decline and then gradual decline. The mixed gas flow rate at the outlet demonstrates an overall trend of rapid increase, followed by slow increase and eventual stabilization, while the CH_4 outflow rate shows a pattern of rapid rise, gradual decline, and final stabilization. The cumulative CH_4 emissions display an initial rapid increase followed by a slower accumulation trend as displacement time progresses.
2. Compared to N_2 injection displacement, CO_2 injection displacement exhibits earlier breakthrough of foreign gas, faster increases in mixed gas flow rate and CH_4 outflow rate at the outlet with higher peak values, as well as greater cumulative CH_4 emissions, due to the combined effects of competitive adsorption and partial pressure between CO_2 and CH_4 in the coal sample container. However, under identical injection gas source conditions, while the growth rate of cumulative CH_4 emissions accelerates with increasing injection flow rates, the total cumulative CH_4 emissions show minimal variation across different flow rates. For all tested injection flow rates, the duration required for CO_2 injection displacement is shorter than that for N_2 injection under equivalent conditions, indicating that CO_2 injection can accelerate CH_4 transport in coal compared to N_2 injection.
3. Compared to constant-rate injection methods, periodic variable-rate injection with maintained low-flow injection of foreign gases can significantly enhance CH_4 production. Under the same periodic injection scheme, smaller differences between peak and trough injection rates result in higher CH_4 emission volumes. When injection duration and average flow rate remain constant, optimized control of foreign gas injection patterns can more effectively promote CH_4 emissions from coal seams, thereby improving CH_4 recovery efficiency. For high-concentration gas utilization purposes, sinusoidal-combined constant-rate injection proves more effective; for low-concentration gas utilization, stepwise injection demonstrates certain advantages; whereas for coal seam gas hazard mitigation, combined sinusoidal, constant-rate and stepwise injection methods are expected to yield optimal results.

4. This study elucidates the differences and connections in CH_4 dynamic transport behavior during gas injection displacement between macroscopic experimental and microscopic pore-scale conditions. Overall, the responses of CH_4 emission volume, displacement duration, and total CH_4 desorption quantity remain consistent across macro-micro scales, where the dynamic transport behavior of CH_4 at the microscopic pore scale determines its macroscopic experimental manifestations. However, since the macroscopic scale simultaneously accounts for desorption of adsorbed CH_4 from both internal and surface regions of the coal matrix, variations emerge in the desorption rate changes of adsorbed CH_4 between the two scales.
5. The simulation curves of the dual-porosity diffusion model for pure CH_4 diffusion in coal demonstrate better agreement with experimental measurements compared to the single-porosity model, indicating that the dual-porosity model more accurately reflects the actual complex pore structure of coal samples. The corresponding diffusion coefficients for macropores and micropores in this model provide a more comprehensive explanation of gas diffusion processes in coal. For CH_4 diffusion during gas injection displacement (whether N_2 or CO_2 injection), comparison of experimentally measured mt/m∞ curves shows that the single-porosity model better fits the experimental curves in early stages, while the dual-porosity model provides closer agreement in later stages, which relates to the diffusion pathways of CH_4 in coal. With increasing injection flow rates, the diffusion parameters in both models show synchronous increases. Moreover, under CO_2 injection displacement, the gas diffusion parameters in both models are lower than those under N_2 injection. Under equivalent pre-adsorption gas pressure conditions, the diffusion coefficients of CH_4 during displacement processes are lower than those during pure CH_4 diffusion for both single-porosity and dual-porosity models, demonstrating that the diffusion-adsorption behavior of foreign gases during injection displacement processes reduces the desorption-diffusion rate of CH_4 in coal to some extent.
6. During CO_2 injection displacement processes, coal matrix adsorption-induced swelling leads to anomalous diffusion phenomena, rendering conventional single-porosity and dual-porosity diffusion models based on Fick's law inadequate for comprehensively describing CH_4 dynamic diffusion behavior under CO_2 injection conditions. Experimental observations reveal that in early stages, these traditional models overestimate diffusion rates compared to measured values, while in later stages they produce underestimations. In contrast, the diffusion-relaxation model demonstrates superior capability in characterizing CH_4 dynamic diffusion behavior throughout CO_2 injection displacement. Under various injection flow rates, the relaxation coefficient (k) consistently exceeds the effective diffusion coefficient (D_e) by 1–2 orders of magnitude, confirming the dominance of relaxation diffusion in governing CH_4 transport during CO_2 injection displacement. Furthermore, both the effective diffusion coefficient (D_e) and relaxation coefficient (k) exhibit positive correlations with increasing injection flow rates throughout the displacement process.

7. Based on the diffusion-relaxation characteristics of CH_4 dynamic transport during the displacement process, the dynamic diffusion of CH_4 under CO_2 injection displacement conditions can be divided into two distinct phases according to the location of "desorption-diffusion" occurrence: In the initial displacement phase, the desorption-diffusion of CH_4 from coal matrix surfaces is primarily driven by concentration gradients, where Fickian diffusion dominates. During the later displacement stage, the desorption-diffusion of CH_4 within the coal matrix interior is governed by relaxation diffusion. The complete dynamic diffusion process of CH_4 under displacement conditions represents the superposition of these two mechanisms.

Chapter 5
Engineering Application of Underground N_2 Injection for Coalbed CH_4 Displacement and Displacement Effect Prediction

5.1 Introduction

The aforementioned experimental results at macroscopic scale demonstrate that injecting CO_2 or N_2 into coal seams can effectively displace the otherwise difficult-to-desorb CH_4, providing experimental evidence that gas injection displacement technology represents an effective approach for gas hazard control. However, since the coal samples used in laboratory experiments were significantly smaller than actual coal masses in underground mines, it is necessary to conduct engineering-scale field applications of underground gas injection displacement for coalbed CH_4 to validate the technology's effectiveness in gas hazard prevention.

Although CO_2 or air injection can also displace coalbed methane, from a long-term perspective, N_2 injection displacement represents a more feasible technology for large-scale implementation in underground coal mines. Comparatively speaking, weakly adsorbed N_2 poses minimal risk of inducing outburst hazards. Moreover, as N_2 is already utilized in underground fire prevention systems, its engineering application offers higher safety than air injection. Therefore, underground N_2 injection displacement technology combines both safety and economic advantages, presenting greater feasibility for widespread engineering implementation.

This chapter employs a self-designed underground "pressure boosting-gas storage-gas injection" test system to investigate the engineering characteristics of CH_4 migration under N_2 injection displacement conditions. The study will: (1) quantitatively analyze the differential effects on CH_4 drainage between gas injection displacement and conventional drainage methods, as well as between intermittent and continuous injection approaches; (2) examine the influence of injection flow rates on CH_4 drainage efficiency; (3) explore the underlying mechanisms of N_2-enhanced CH_4 drainage at engineering scale based on application results; and (4) conduct preliminary exploration and model optimization for applying deep learning algorithms in predicting gas injection displacement effectiveness at engineering scale.

G. Wang, H. Gong, *Dynamic Migration Characteristics of CH4 During Gas Injection Displacement in Coal Seams*,
https://doi.org/10.1007/978-981-95-8879-4_5

5.2 Overview of the Test Mine

Liuzhuang Coal Mine is located in Fuyang City, Anhui Province, the construction of which started in February 2003 with a designed annual production capacity of 8.0 Mt./a. In June 2012, Anhui Economic and Information Technology Commission and Anhui Coal Mine Safety Supervision Bureau issued No. 608 of Anhui Economic and Information Technology Coal Letter〔2012〕, approving the production capacity of Liuzhuang Coal Mine as 11.4 Mt./a. As one of the main production mines of China Coal Xincheng Energy Company Limited, Liuzhuang Coal Mine was recognized as a coal and gas prominent mine in 2017. The main coal seam $11\text{–}2^{\#}$ was selected as the non-protruding coal seam, with a thickness of 1.37–5.40 m and an average of 3.29 m. The lithology of the top plate of this coal seam is mainly sandy mudstone and mudstone, followed by sandstone; the lithology of the bottom plate is mainly sandy mudstone and mudstone, followed by sandstone. The measured gas content is between 1.90 and 4.33 m^3/t, which is low, and the measured gas pressure is 0.38 MPa.

Historical gas parameter statistics from Liuzhuang Coal Mine reveal that despite the seam's low gas content and pressure, the mining faces experience significant gas emissions during extraction due to high mining intensity. The average monthly frequency of gas concentrations exceeding 0.6% at upper corners reaches seven occurrences, presenting substantial gas control challenges. To enhance targeted gas prevention, the mine has implemented various technical measures, including hydraulic permeability enhancement, high-pressure air blasting, ultra-high pressure hydraulic slotting, CO_2 phase-transition fracturing, and controlled shock wave techniques, all with limited effectiveness.

According to Darcy's seepage theory, gas flow velocity exhibits positive correlations with both pressure gradient and permeability. Coal seam gas content and pressure collectively influence gas flow rates during drainage operations. The low gas content and pressure characteristics of Liuzhuang's coal seams inherently limit conventional drainage effectiveness. While permeability enhancement technologies may show short-term efficacy, the seam's inherent properties render traditional modification techniques inadequate for sustained improvement in gas control performance.

5.3 Engineering Overview

5.3.1 Test Site and Borehole Layout

The experimental site was located at borehole sites $10^{\#}$, $13^{\#}$, and $16^{\#}$ along the belt entry of panel 151,105 in the $11\text{–}2^{\#}$ coal seam of Liuzhuang Mine's western section. For this engineering practice, a total of six sets of in-seam boreholes were deployed. To ensure gas tightness of experimental boreholes, all were sealed using the "two-plug-one-injection" method with an expansion cement material at a sealing depth of 15 m. Post-sealing integrity tests were conducted, with secondary grouting implemented where leakage was detected. The geographical location of Liuzhuang Mine's western section, spatial relationships between injection/production boreholes, borehole distribution patterns, and detailed borehole parameter designs are illustrated in Fig. 5.1.

5.3.2 Construction of the "Pressurization-Gas Storage-Gas Injection" Experimental System

To ensure the smooth implementation of field engineering applications, a "pressure boosting-gas storage-gas injection" experimental system was established utilizing the inherent conditions and characteristics of the underground coal mine. The

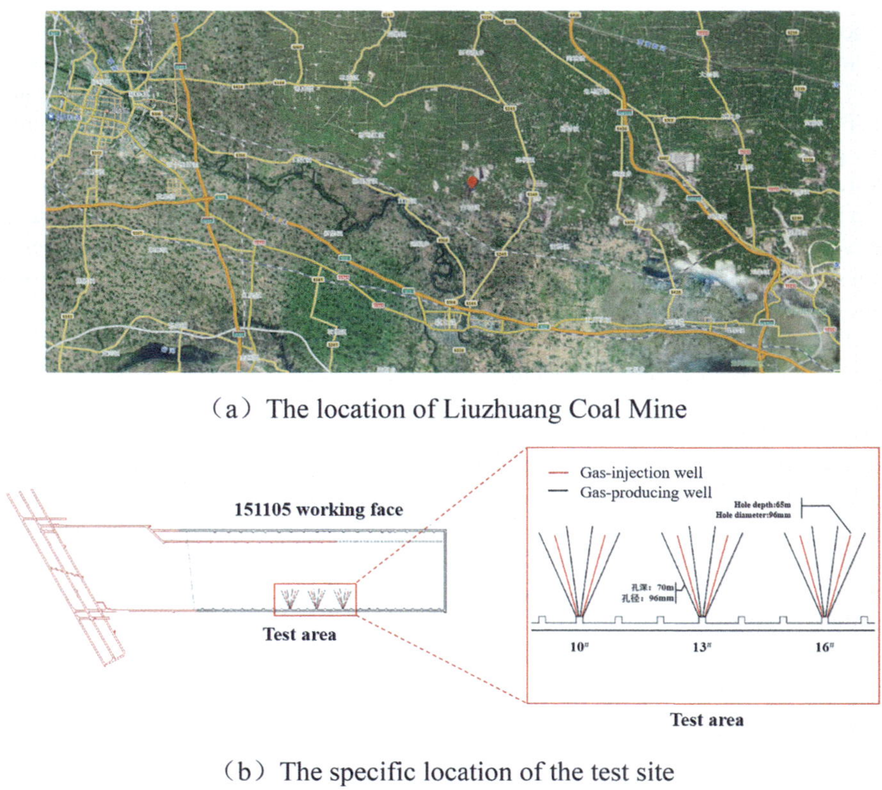

(a) The location of Liuzhuang Coal Mine

(b) The specific location of the test site

Fig. 5.1 Location of Liuzhuang Coal Mine (**a**) and the micro-pilot test in 151,105 working face (**b**). (**a**) The location of Liuzhuang Coal Mine. (**b**) The specific location of the test site

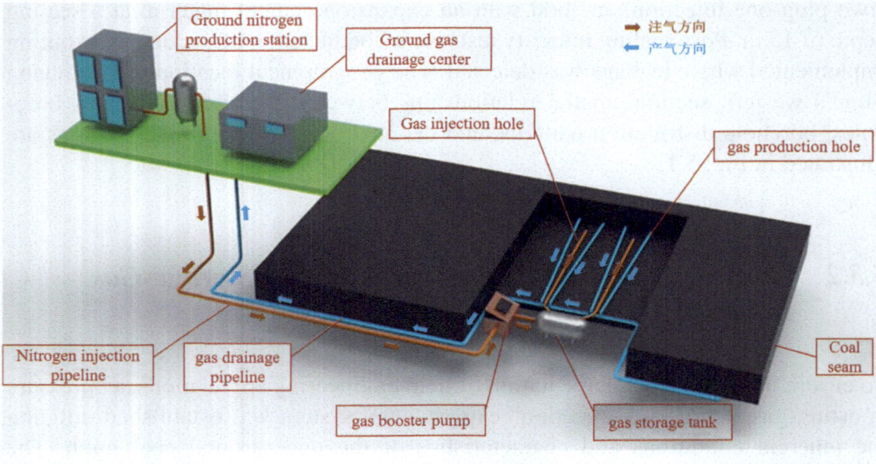

Fig. 5.2 Principle of N_2 injection replacement field tests [220]

schematic diagram of the experimental system is shown in Fig. 5.2, while the equipment used for the field N_2 injection displacement test system is presented in Fig. 5.3.

The experimental system consists of five subsystems: (1) N_2 production and delivery system, (2) pneumatic pressure boosting system, (3) gas injection system, (4) gas production system, and (5) gas analysis system. The N_2 production and delivery system includes surface N_2 generators, underground N_2 delivery pipelines, and pressure-resistant gas delivery hoses. The pneumatic pressure boosting system is equipped with pneumatic booster pumps, high-pressure gas delivery hoses, and gas storage tanks, where the booster pumps increase injection gas pressure and the storage tanks contain high-pressure gas.

The gas injection system comprises metal tube float flowmeters, gas injection pressure regulating valves, injection steel pipes, flame-retardant antistatic PVC sealing pipes, and flame-retardant antistatic PVC injection screen pipes. The metal tube float flowmeters monitor injection flow rates, while the pressure regulating valves control injection pressure. The injection steel pipes, sealing pipes, and screen pipes collectively form the N_2 injection pathway within boreholes. The gas production and analysis system includes wet gas flowmeters, gas sampling bags, and gas chromatographs. Wet gas flowmeters measure gas flow rates in pipelines, gas sampling bags collect gas samples from production boreholes, and gas chromatographs analyze the composition of mixed gases from production boreholes.

5.3.3 Experimental Design and Procedure

The experimental scheme for N_2 injection displacement to enhance coalbed CH_4 drainage is presented in Table 5.1. The test was conducted at three locations, with two sets of boreholes drilled at each site following a "one-injection-two-production"

Fig. 5.3 Equipment used in the N_2 injection replacement field test system. (**a**) N_2 production and conveying system. (**b**) Pneumatic pressurization system. (**c**) Gas injection system. (**d**) Gas production analysis system

Table 5.1 N_2 injection to promote the discharge of coal seam CH_4 test program

Test position	Injection borehole	Production borehole	Gas injection flow rate
10#Drilling Site	IB-1, IB-2	PB1–1, PB1–2, PB2–1, PB2–2	10 m^3/h
13#Drilling Site	IB-3, IB-4	PB3–1, PB3–2, PB4–1, PB4–2	15 m^3/h
16#Drilling Site	IB-5, IB-6	PB5–1, PB5–2, PB6–1, PB6–2	20 m^3/h

configuration. The implementation adhered to an "initial baseline monitoring followed by gas displacement" sequence, establishing data benchmarks through preliminary parameter collection for subsequent displacement effectiveness evaluation. The N_2 injection displacement operation was divided into distinct phases:

Phase I: Conventional CH_4 drainage monitoring (May 14–July 10, 58 days) established natural emission baselines.

Phase II: Primary N_2 injection (July 11–24, 14 days) implemented initial displacement.

Phase III: Injection suspension observation (July 25–30, 6 days) analyzed post--displacement effects.

Phase IV: Secondary N_2 injection (August 1–11, 10 days) evaluated enhanced displacement efficiency.

To investigate injection flow rate effects on CH_4 drainage enhancement, control group tests were conducted at 10, 15, and 20 m^3/h, respectively.

5.4 Field Test Results and Analysis

5.4.1 Parameter Variations in Production Wells

Taking the field test of N_2 injection for coalbed CH_4 displacement at Drilling Site $10^{\#}$ as an example, this section investigates the engineering characteristics of parameter variations in production wells during the N_2 injection and displacement process.

5.4.1.1 Gas Composition Variations in Production Wells

This experiment employed gas chromatography (GC) analysis to quantitatively characterize the chemical composition of mixed gases discharged from four production boreholes. Approximately 50 ml gas samples were extracted from each borehole and introduced into the chromatograph through the laboratory instrument's sample introduction system. To ensure data reliability, all samples underwent duplicate parallel testing. The final analytical results revealed the gas composition included methane (CH_4), nitrogen (N_2), oxygen (O_2), as well as various compounds, such as carbon dioxide (CO_2), carbon monoxide (CO), and ethane (C_2H_6). This study focuses exclusively on analyzing the variations in methane and nitrogen components.

Figure 5.4 illustrates the systematic variations in CH_4 and N_2 composition distributions across four production boreholes throughout the experimental cycle. During conventional drainage, the CH_4 volume fraction exhibited a fluctuating decline from an initial concentration of approximately 75%, eventually stabilizing within the 55% range. In contrast, the N_2 component demonstrated inverse concentration trends, with its volume fraction progressively increasing from a baseline of about 20% to a stable range near 30%. This differential behavior stems from coal's stronger adsorption affinity for CH_4 compared to N_2, resulting in faster N_2 desorption

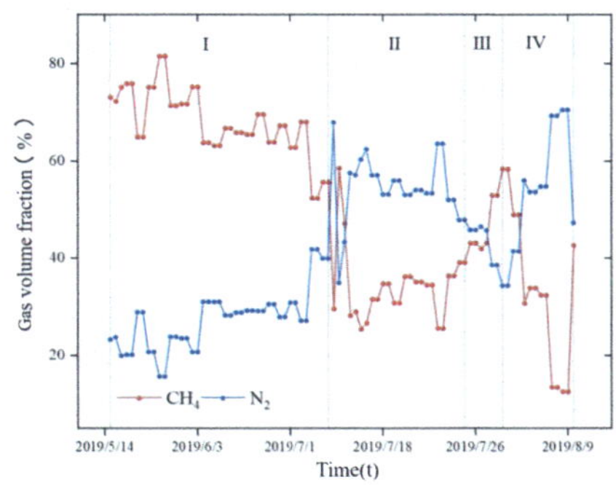

(a) Borehole P1-1

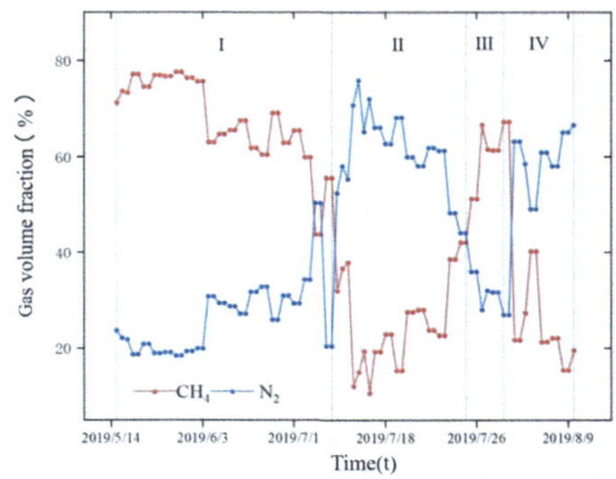

(b) Borehole P1-2

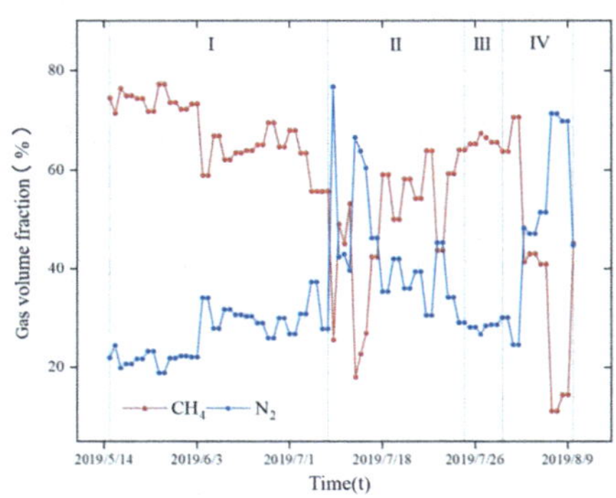

(c) Borehole P2-1

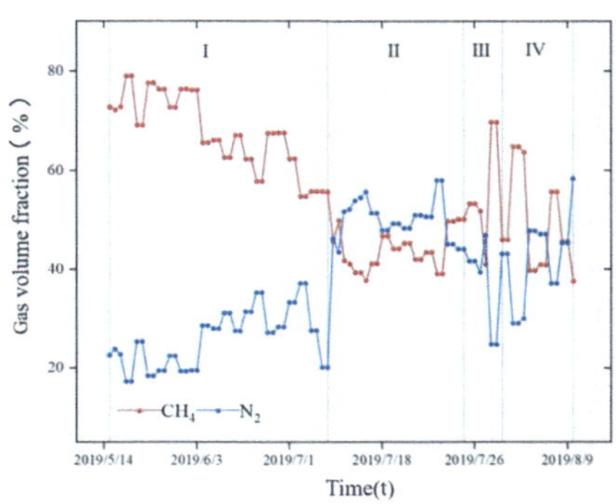

(d) Borehole P2-2

Fig. 5.4 CH_4 and N_2 composition in each gas-production borehole. (**a**) Borehole P1-1. (**b**) Borehole P1-2. (**c**) Borehole P2-1. (**d**) Borehole P2-2 rates during conventional drainage that drive the observed opposing concentration trends.

The initial injection phase revealed distinct compositional dynamics. All four production boreholes showed rapid N_2 volume fraction increases (200–300% enhancement) followed by gradual declines, a phenomenon attributable to the interconnected macro-fracture networks formed during borehole development in the fan-shaped spatial distribution. Concurrently, CH_4 volume fractions experienced abrupt decreases initially, with subsequent recovery patterns varying significantly among boreholes. Secondary injection periods maintained an overall downward CH_4 volume fraction trend while N_2 concentrations continued rising. Notably, during injection suspension intervals, CH_4 volume fractions rebounded to mid-phase conventional drainage levels. This characteristic behavior originates from two persistent mechanisms: first, substantial concentration gradients of adsorbed CH_4 between fracture systems and coal matrices maintain high desorption rates despite ceased N_2 injection; second, sustained pressure gradients continue driving directional CH_4 migration from reservoir fractures into borehole spaces.

5.4.1.2 Mixed Gas Flow Rate Variations in Production Wells

The variations in mixed gas flow rates at each production well are shown in Fig. 5.5.

During the conventional drainage phase, all production boreholes exhibited consistently declining gas flow rates. This phenomenon occurs because the gas pressure within boreholes penetrating the coal seam remains relatively low, establishing a pressure gradient (pressure differential) between the boreholes and surrounding coal mass. This gradient drives the flow of free-phase CH_4 from inter-block fractures toward production boreholes. As drainage progresses, the gas pressure within coal fractures gradually decreases, disrupting the dynamic equilibrium between adsorbed and free-phase CH_4. Consequently, adsorbed CH_4 desorbs from the coal matrix and diffuses into inter-block fractures, replenishing the free-phase CH_4 (serving as a gas source). With continued drainage, the total CH_4 content in the coal seam shows a progressive decline, while the pressure gradient between the borehole–coal system undergoes concomitant weakening. This results in attenuated gas migration driving forces and subsequent reduction in mixed gas flow rates at production boreholes. Data presented in Fig. 5.5 indicate that by the end of conventional drainage, the mixed gas flux at respective boreholes had decreased to 5.22e-4, 4.00e-4, 6.25e-4, and 6.38e-4 m^3/min.

During the gas injection implementation phase, production parameters exhibited distinct gas transport enhancement characteristics. Both primary and secondary injection phases demonstrated rapid escalation in mixed gas flow rates at production boreholes, which, after peaking, maintained elevated levels despite subsequent declines. This phenomenon is attributed to the sustained pressure gradient driving gas migration through fractures. As evidenced by Fig. 5.5, the average mixed flow rates during primary injection, injection suspension, and secondary injection phases increased by factors of 2.56, 5.83, 2.97, and 2.20, respectively, compared to

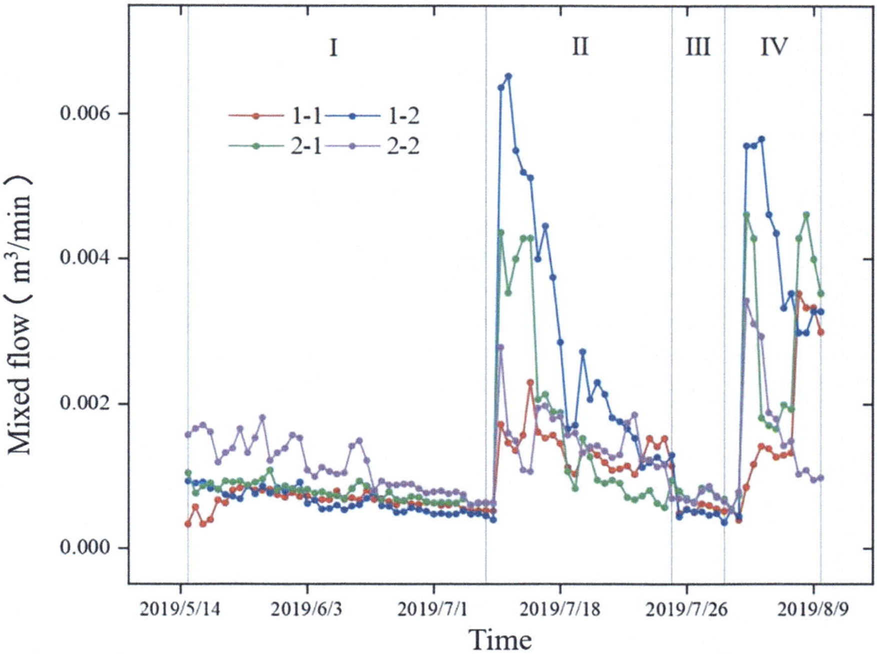

Fig. 5.5 Variations in the flow rates of mixed gases from each gas-production borehole

conventional drainage termination values. Notably, a progressive attenuation of mixed gas flow rates was consistently observed during continuous injection periods, primarily caused by decreasing injection velocities resulting from water accumulation in pipelines over extended operation. This led to gradual decay of pressure gradients between injection and production boreholes. During suspension intervals, mixed gas flow rates displayed intermediate characteristics—lower than active injection phases yet significantly higher than conventional drainage values. This transitional state empirically validates the feasibility of intermittent injection technology. Throughout intermittent injection and post-termination observation periods, all production boreholes maintained both instantaneous and average mixed flow rates exceeding pre-injection baselines. These findings confirm that pressurized gas injection enhances pressure gradients between coal fractures and production boreholes, effectively promoting gas flow through the fracture network and demonstrating clear "flow enhancement" effects.

5.4.1.3 Coalbed CH_4 Content Variations Pre- and Post-Displacement

To evaluate the effectiveness of the nitrogen injection displacement process, coal seam gas content measurements were conducted both before and after the operation within the test area. The distribution of measurement points is illustrated in Fig. 5.6, with corresponding results presented in Table 5.2.

The measurement data demonstrate a significant reduction in coal seam gas content following nitrogen injection displacement. Specifically, the average gas content in the affected zone of Injection Borehole 1 decreased by 1.0551 m^3/t (representing a 25.31% reduction), while the affected zone of Injection Borehole 2 showed a decrease of 1.2469 m^3/t (corresponding to a 28.05% reduction). These results confirm that N_2 injection displacement technology effectively reduces coal seam gas content, thereby mitigating the potential for gas-related hazards.

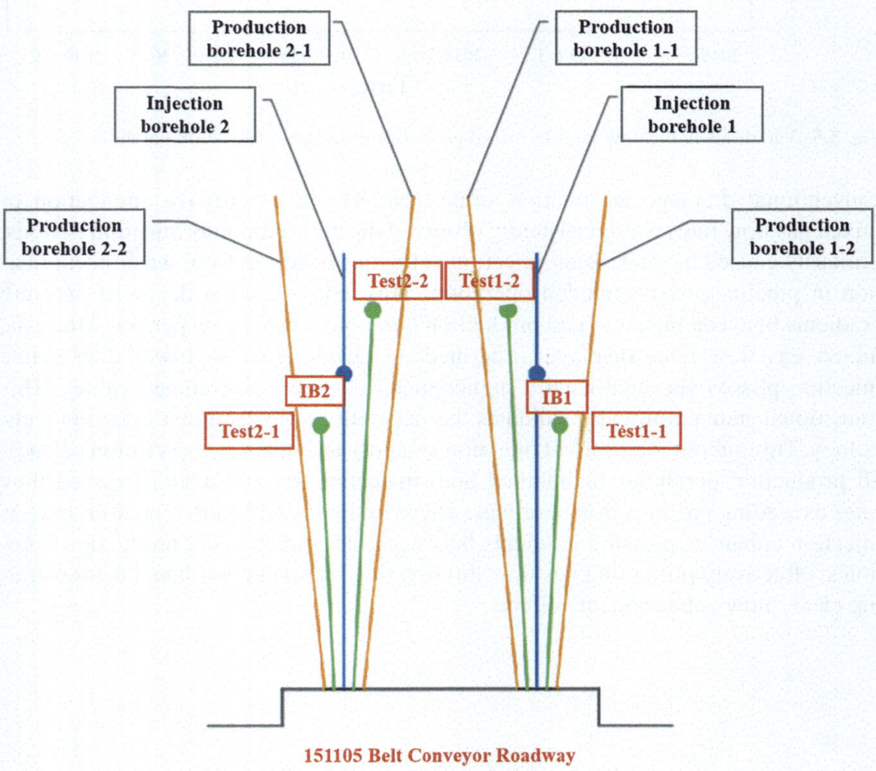

Fig. 5.6 Distribution of sampling points

Table 5.2 Comparison of coal seam gas content before and after gas injection

Pre-injection		Post-injection		
Sample ID	Coalbed methane content (m^3/t)	Sample ID	Coalbed methane content (m^3/t)	Mean value (m^3/t)
IN1	4.1695	Test1–1	2.9653	3.1144
		Test1–2	3.2634	
IN2	4.4450	Test2–1	3.1854	3.1981
		Test2–2	3.2108	

5.4.1.4 CH_4 Pure Flow Rate Variations in Production Wells

The pure flow rates of CH_4 and N_2 (Q_i) for each production borehole were calculated using the following equation:

$$Q_i = Q_M \times C_i \tag{5.1}$$

In the equation, Q_i represents the outlet flow rate of each gas component from the gas production borehole (m^3/min); Q_M denotes the mixed gas flow rate from the gas production borehole (m^3/min); C_i indicates the gas composition of each component in the gas production borehole (%); and i refers to CH_4 and N_2.

As illustrated in Fig. 5.7, during conventional drainage, the pure CH_4 flow rates from four production boreholes exhibited an initial slight increase, peaked around day 10, and then entered a gradual decline phase, with CH_4 flow rates consistently exceeding those of N_2 from the same boreholes. Following injection initiation, both CH_4 and N_2 pure flow rates showed significant surges: during the first injection phase, CH_4 flow rate increases reached 76.37%, 818.39%, 221.43%, and 258.26%, while the second injection phase produced increases of 79.84%, 299.71%, 590.51%, and 511.19%. Despite decreasing CH_4 concentrations during injection, the absolute flow rates remained elevated. Experimental data revealed that average CH_4 flow rates after initial injection commencement exceeded terminal conventional drainage values by factors of 1.42, 2.66, 2.21, and 1.89 for respective boreholes.

Notably, N_2 flow rates exceeded CH_4 in the first borehole group during both injection phases, while the second group showed no consistent flow rate dominance pattern. During injection suspension phases, both CH_4 and N_2 flow rates declined, though CH_4 rates remained higher than both concurrent N_2 rates and post-conventional drainage CH_4 values. Flow rates resumed growth upon reinjection. Across all four phases, CH_4 flow rates universally demonstrated initial increase followed by decrease, whereas N_2 exhibited this pattern only during active injection phases, showing gradual decline during suspension and minimal variation during conventional drainage.

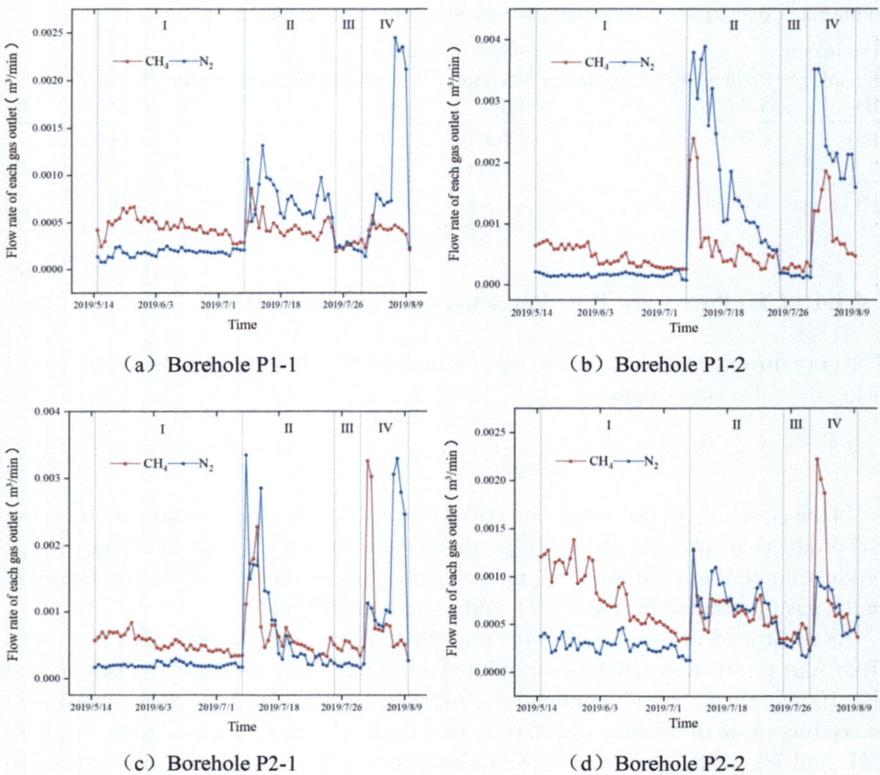

Fig. 5.7 CH_4 and N_2 outlet flow rate of each gas-producing borehole. **(a)** Borehole P1-1. **(b)** Borehole P1-2. **(c)** Borehole P2-1. **(d)** Borehole P2-2

5.4.2 Comparison of CH_4 Extraction Efficiency Between Conventional Drainage and Gas Injection Displacement Processes

The ultimate objective of applying underground displacement technology is to enhance coalbed methane production. The achievement of production enhancement and its magnitude serve as direct validation of the effectiveness of underground N_2 injection displacement technology, thereby providing critical evidence for evaluating its potential for large-scale implementation. In this study, we define superior coalbed gas drainage performance as corresponding to greater cumulative CH_4 emissions from production boreholes within equivalent timeframes.

This section conducts fitting and nonlinear prediction of CH_4 outflow rate variation curves during conventional drainage phases. By comparing the predicted cumulative CH_4 emissions under conventional drainage conditions with measured cumulative emissions from field gas injection displacement tests, we perform comprehensive effectiveness analysis of the practical application of underground N_2 injection displacement technology.

Figure 5.8 presents the predicted curves for the conventional drainage phase of each production borehole in this field test. The CH_4 outflow rates from all production boreholes exhibited an initial increase followed by gradual attenuation, a process that can be well characterized by second-order exponential decay functions. Following the initiation of N_2 injection displacement tests, the majority of measured CH_4 outflow data points from production boreholes were observed above the predicted curves. This demonstrates that the application of N_2 injection displacement technology can enhance CH_4 flow rates compared to conventional drainage. However, due to uncertainties in field conditions, the degree of CH_4 flow rate enhancement varied across different measurements.

This section conducts data analysis by comparing the cumulative CH_4 emissions under simultaneous injection-production conditions with the predicted cumulative CH_4 emissions calculated using the conventional drainage fitting formula for the corresponding period. The cumulative CH_4 emissions for each borehole are presented in Fig. 5.9.

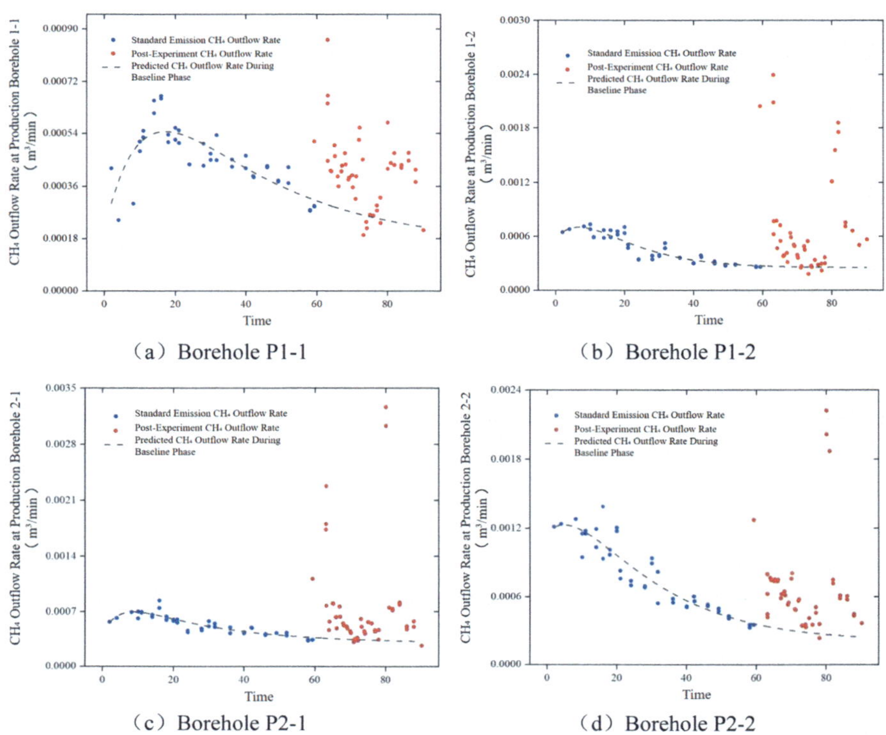

(a) Borehole P1-1 (b) Borehole P1-2

(c) Borehole P2-1 (d) Borehole P2-2

Fig. 5.8 Predicted CH_4 export fluxes during the conventional emission phase for each gas--producing borehole. (**a**) Borehole P1-1. (**b**) Borehole P1-2. (**c**) Borehole P2-1. (**d**) Borehole P2-2

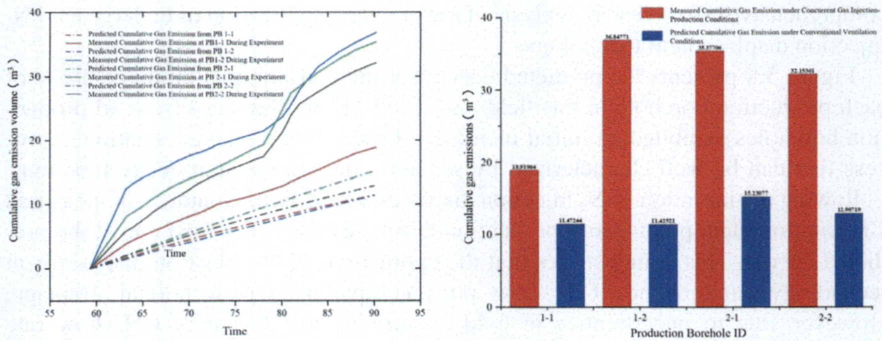

(a) The cumulative CH_4 emission curve (b) Total Cumulative CH_4 Emissions

Fig. 5.9 Cumulative gas emissions from boreholes under conventional emission projections and "injection and discharge" conditions. (**a**) The cumulative CH_4 emission curve. (**b**) Total cumulative CH_4 emissions

As shown in Fig. 5.9a, throughout the entire gas injection test period, the growth rate of cumulative CH_4 emissions under simultaneous injection-production conditions at each production borehole significantly exceeded the predicted growth rate under conventional drainage conditions. Notably, during the initial stages of both primary and secondary injection phases, the slope of the cumulative CH_4 emission curve under simultaneous injection-production conditions was markedly steeper than during the middle and later stages of the injection phases. This field observation provides empirical support for the feasibility of intermittent or cyclic injection approaches.

Figure 5.9b reveals that the total cumulative CH_4 emissions from production boreholes 1-1, 1-2, 2-1, and 2-2 under simultaneous injection-production conditions were 18.91964, 36.84771, 35.37706, and 32.15301 m^3, respectively. In contrast, the predicted cumulative emissions under conventional conditions were 11.47244, 11.42521, 15.23077, and 12.90719 m^3, respectively. This represents percentage increases of 64.9%, 222.5%, 132.3%, and 149.1% in total CH_4 emissions under simultaneous injection-production conditions compared to the predicted values for corresponding production boreholes.

5.4.3 *Comparative Analysis of CH_4 Drainage Performance Between Continuous and Intermittent Injection Methods*

The CH_4 outflow rates during the primary injection phase were analyzed through nonlinear regression, with subsequent trend predictions generated from the fitted function curves, as presented in Fig. 5.10. Analysis reveals that during the

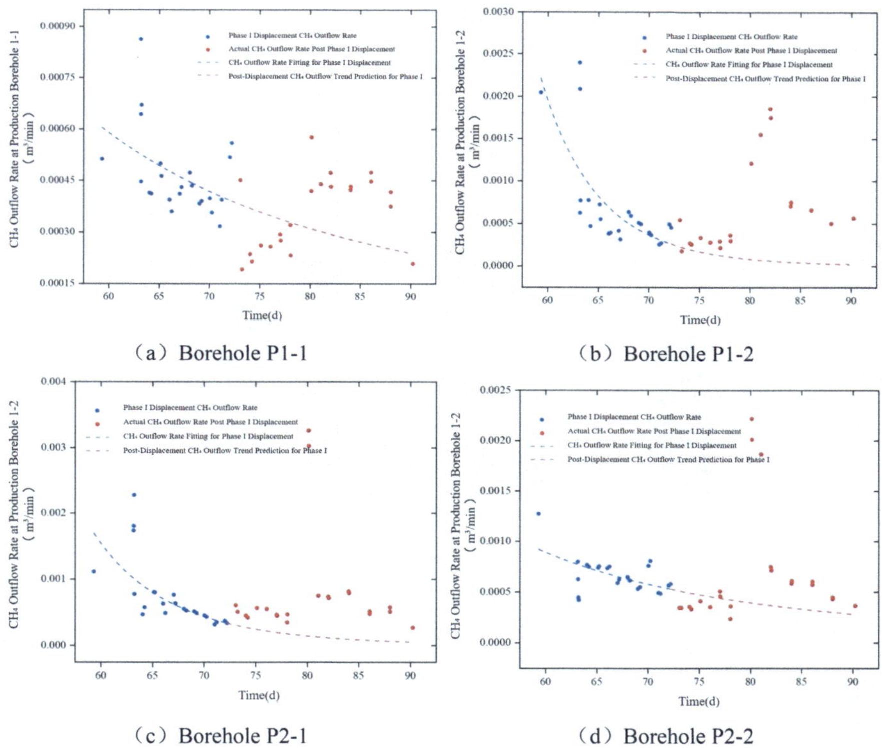

Fig. 5.10 Predicted CH_4 outlet flow from continuous gas injection at each gas-producing borehole. (**a**) Borehole P1-1. (**b**) Borehole P1-2. (**c**) Borehole P2-1. (**d**) Borehole P2-2

secondary injection phase, the measured pure CH_4 flow rates at all production boreholes consistently exceeded the predicted values for continuous injection.

This phenomenon stems from the dynamic processes occurring during the injection suspension period. The progressive reduction of pressure gradients between the injection-affected coal matrix and production boreholes leads to declining mixed gas flow rates. Simultaneously, in the absence of N_2 replenishment, the increasing CH_4 concentration results in substantial CH_4 accumulation within coal fracture networks.

Following injection resumption, the reestablished pressure gradient facilitates rapid discharge of the accumulated CH_4 through production boreholes, manifesting as a characteristic flow rate surge. While the CH_4 outflow rates exhibit a declining trend similar to the primary injection phase, they maintain consistently higher values than the continuous injection predictions throughout the subsequent injection period.

As clearly shown in Fig. 5.11, the comparative analysis demonstrates that during the period from the end of initial gas injection to test termination, the estimated cumulative CH_4 emissions from each production borehole under continuous

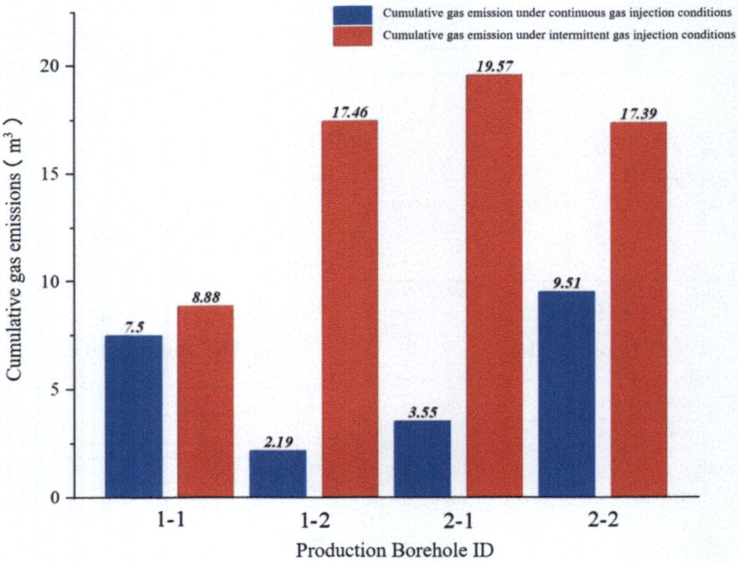

Fig. 5.11 Cumulative gas emissions by borehole for continuous injection predictions vs. intermittent injection conditions

injection were 7.5, 2.19, 3.55, and 9.51 m^3, respectively. In contrast, intermittent injection yielded significantly higher cumulative CH_4 outputs of 8.88, 17.46, 19.57, and 17.39 m^3 from corresponding boreholes, indicating markedly superior CH_4 drainage enhancement compared to continuous injection. This evidence confirms that periodic injection represents the preferable approach for both coalbed methane production enhancement and gas hazard mitigation. The cyclical injection method offers dual advantages: first, it enables substantial reduction in gas injection volume through optimized injection start-stop control, thereby lowering operational costs; second, it achieves greater CH_4 production within equivalent timeframes, significantly improving economic returns.

5.4.4 Comparative Analysis of CH_4 Drainage Enhancement Under Different Gas Injection Flow Rates

As indicated in Sect. 5.4.1, there exist certain variations in gas production parameters among different production boreholes under identical conditions. To minimize the impact of these variations when comparing the CH_4 drainage effects under different gas injection flow rates, this section analyzes the gas production parameters collected at the manifolds of each drilling site.

Figure 5.12 shows the variations in pure CH_4 flow rates at the drilling site manifolds under different injection flow rates. The results indicate that during both injection phases, higher injection flow rates lead to faster initial increases in pure CH_4 flow rates and higher peak values. This occurs because greater injection flow rates create larger pressure gradients between the major fractures in the coal mass and the production boreholes, resulting in more efficient displacement of free-phase CH_4 per unit time.

During the primary injection phase, the peak pure CH_4 flow rates at the return manifolds were 0.00581, 0.00863, and 0.01006 m^3/min under injection flow rates of 10, 15, and 20 m^3/h, respectively. In the secondary injection phase, the corresponding peak pure CH_4 flow rates were 0.00713, 0.01259, and 0.01591 m^3/min. However, the duration of this effect decreases with increasing injection flow rates. As the gas injection displacement progresses, higher injection flow rates result in faster declines in pure CH_4 flow rates and greater reduction magnitudes. During the later stages of gas injection displacement, the differences in pure CH_4 flow rates under various injection flow rates gradually diminish.

The pure CH_4 flow rates during primary injection, injection suspension, and secondary injection phases were measured to calculate the final cumulative CH_4 emissions at the manifold of each drilling site under different injection flow rates, as illustrated in Fig. 5.13. The results demonstrate minimal variation in cumulative

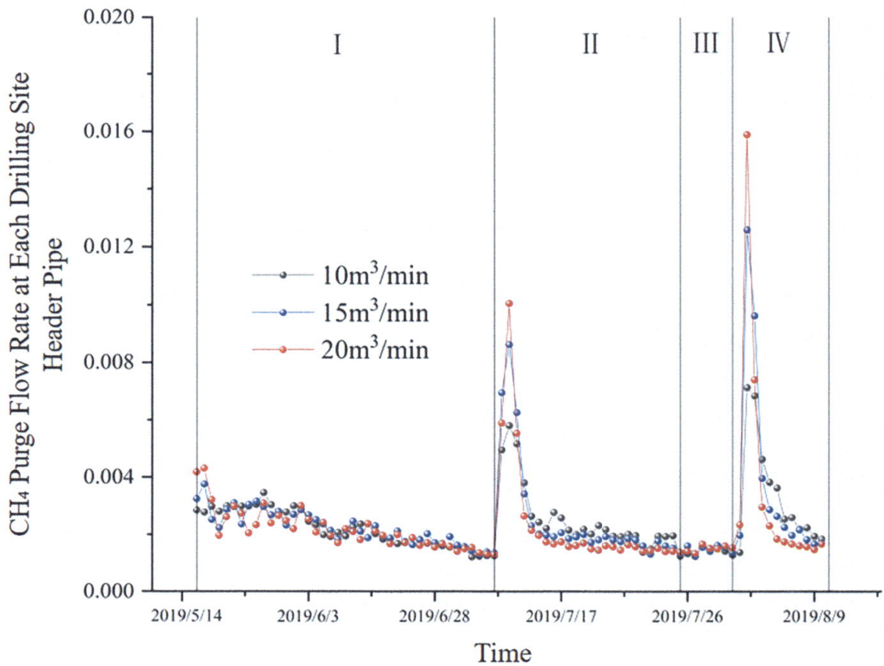

Fig. 5.12 Variation of CH_4 pure flow rate at the drilling site manifold for each gas injection flow rate

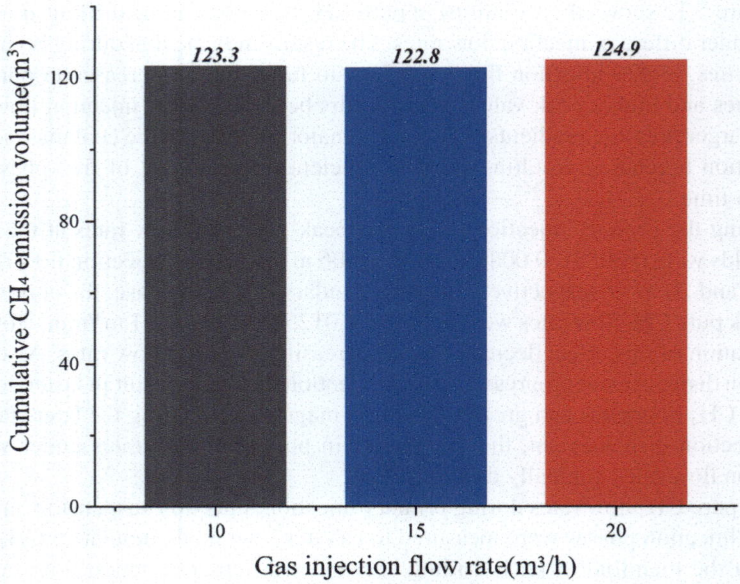

Fig. 5.13 Cumulative final CH_4 emissions at the drillfield confluence for each injection flow rate

CH_4 emissions across different injection flow rates. Specifically, at injection flow rates of 10, 15, and 20 m^3/h, the respective final cumulative CH_4 emissions at the manifolds were 123.3, 122.8, and 124.9 m^3. This suggests that while injection flow rate affects the CH_4 transport velocity within the coal mass at engineering scale, it has negligible impact on the cumulative CH_4 emissions—a finding consistent with the conclusions drawn in Sects. 4.4.1 and 3.4.2.

5.5 CH_4 Migration-Production Mechanism During N_2 Stimulation at Engineering Scale

It is widely acknowledged that coal seams constitute a dual-porosity system comprising both inter-block fractures and coal matrix pores [221]. Gas flow within inter-block fractures adheres to Darcy's law, while gas transport through coal matrix pores follows Fick's diffusion law. Building upon the two-phase gas injection displacement mechanism established through laboratory experiments in Ref. [3], the underground N_2 injection coalbed CH_4 displacement process can be divided into two distinct phases: Injection Phase I and Injection Phase II (Fig. 5.14).

During the conventional drainage phase, the gas pressure at production boreholes (P_0) approximates atmospheric pressure, while the internal CH_4 pressure within the coal mass remains higher than the borehole gas pressure ($P_{f1} > P_0$). Under these conditions, gas exchange between these regions is predominantly governed by

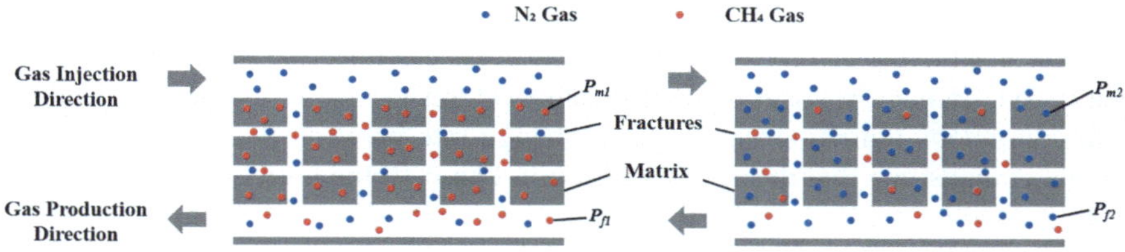

Fig. 5.14 Schematic diagram of N_2 injection to drive the discharge of coal seam CH_4

Darcy flow. The existing pressure gradient (ΔP) drives CH_4 migration from the coal matrix through fractures into production boreholes. As drainage progresses, the pressure gradient (ΔP) gradually diminishes, resulting in CH_4 outflow rates at production boreholes following a second-order exponential decay function.

Prior to N_2 injection into the coal mass, the gas pressures between the matrix and fractures maintain a quasi-dynamic equilibrium state ($P_m = P_f$). Following the initiation of gas injection (Injection Phase I), the total gas pressure within the coal mass ($P = P_f + P_m$) increases and stabilizes, consequently enhancing the pressure differential (ΔP) between the coal mass and production boreholes across fractures. According to Darcy's law, the gas seepage velocity (v) within the coal mass can be expressed as:

$$v = \frac{K}{\mu}\frac{\Delta P}{\Delta L} \tag{5.2}$$

Herein, K denotes the permeability of coal matrix; μ represents gas viscosity; and ΔL stands for fracture length. It can be observed that the greater the pressure difference ΔP between coal matrix and drainage borehole fractures, the faster the gas seepage velocity within the fractures of coal matrix. Experimental results demonstrate that during the initial stage of gas injection, both the mixed gas outlet flow rate and CH_4 outlet flow rate at production boreholes are significantly higher compared to conventional drainage stages. Meanwhile, during Gas Injection Phase I, primarily original free-state CH_4 serves as the main product.

During Gas Injection Phase II, adsorbed CH_4 molecules on the surface of coal matrix pores desorb to supplement CH_4 production at boreholes. With continuous N_2 injection and the constant flushing out of free-state CH_4 in fractures, a CH_4 concentration gradient forms between the matrix and fractures, providing the driving force for gas exchange between them. Consequently, adsorbed CH_4 desorbs and diffuses into fractures, while foreign gases (N_2) migrate in the opposite direction. This gas exchange process can be described by Fick's diffusion la [223]:

$$Q_i = D_i M_i \nabla C_{f_i - m_i} \tag{5.3}$$

In this context, Q_i represents the gas exchange mass between the matrix and fractures; $\nabla C_{fi\text{-}mi}$ denotes the concentration gradient of different gases between the matrix and fractures; D_i stands for the diffusion coefficient of different gases; and M_i indicates the molar mass of different gases. As gas injection proceeds, the concentration gradient between the matrix and fractures gradually diminishes, leading to a continuous reduction in the gas exchange rate between them. Experimental results demonstrate that during Gas Injection Phase II, the outlet flow rate of CH_4 from production pores exhibits an overall declining trend with prolonged injection time. During this phase, CH_4 desorbed from the adsorbed state serves as the primary product.

Furthermore, coal seam permeability is fundamentally a dynamically varying parameter jointly determined by coal fracture stiffness, in situ stress, and coal seam gas pressure [60, 222]. Specifically, external stress acts to compress the coal matrix, thereby reducing permeability, while CH_4 pressure functions to expand the coal matrix, consequently enhancing permeability. The injection of high-pressure gas into coal seams not only promotes gas flow but also reduces the effective stress acting on coal pores, ultimately improving coal seam permeability.

5.6 Application of Deep Learning Algorithms in Displacement Effect Prediction

The CO_2/N_2-ECBM process represents a complex physical phenomenon under multi-field coupling conditions. Current mainstream ECBM simulators exhibit varying degrees of the following issues during displacement simulation and prediction:

(a) While localized comparisons between measured and simulated results demonstrate good agreement, significant discrepancies emerge when examining the complete production cycle;
(b) As fundamentally numerical simulation software, the physical-mathematical models employed by these simulators may not be fully applicable to actual field engineering conditions;
(c) The complexity of real-world engineering conditions makes it challenging to comprehensively incorporate all appropriate physical-mathematical models, and excessive model inclusion may lead to simulation non-convergence.

In recent years, with advancements in big data and artificial intelligence, machine learning has emerged as a nonlinear prediction method based on autonomous machine learning. By circumventing the need to consider the complex physical processes of gas injection and displacement, while demonstrating strong computational and learning capabilities, it has found extensive applications in the complex geological conditions of underground coal mines, including coal and gas outburst prediction [224], coal-rock strength assessment [225], and spontaneous combustion forecasting in goaf area [226]. However, from the perspective of coal seam gas injection and displacement technology, research on displacement effect prediction using deep learning algorithms remains scarcely reported. In this section, machine learning algorithms were employed to predict gas injection and displacement effects, with model applicability analyzed to identify the optimal prediction model, thereby providing references for the application of machine learning algorithm models in predicting underground gas injection and displacement effects.

5.6.1 *Selection of Deep Learning Algorithms and Algorithmic Workflow*

5.6.1.1 BP Neural Network Model

The learning process of the BP neural network model [227, 228] consists of forward signal propagation and backward error propagation. During forward propagation, the model operates on the input layer. After processing through hidden layers, the signal enters the backward propagation phase where errors propagate from the output layer back to the hidden layers. Subsequently, the system transmits the errors back to the input layer in specific substructure forms, distributing them across all units layer by layer, thereby generating reference errors or error signals for each layer. These signals serve as the basis for adjusting the weights of each unit. The continuous optimization of weights constitutes the core mechanism of neural network learning. This process continues iteratively until the network output error gradually reduces to an acceptable range or until the predetermined number of learning iterations is reached. The overall structure of the model is illustrated in Fig. 5.15. The specific implementation process of this algorithm is as follows [229]:

1. Model Node Determination: The input parameters of the model include gas injection pressure, mixed gas flow rate, and gas concentration at production boreholes. The CH_4 flow rate at production boreholes serves as the output parameter of the model. Based on the selected input and output parameters, the number of nodes in the input layer is 3, while the output layer contains 1 node. The number of nodes in the hidden layer is determined through trial-and-error method, as shown in Eq. (5.4).

$$m = \sqrt{n + l} + \alpha \tag{5.4}$$

   In the equation, m represents the number of hidden layer nodes; n denotes the number of input layer nodes; l indicates the number of output layer nodes; and α is a constant between 1 and 10.
2. Initialize the connection weights W_{ij} between input-hidden layers and W_{jk} between hidden-output layers, respectively. Initialize the hidden layer threshold a and output layer threshold b.
3. The data transfer in the model is shown in Eqs. (5.5)–(5.12):

   (a) Calculation of hidden layer output:

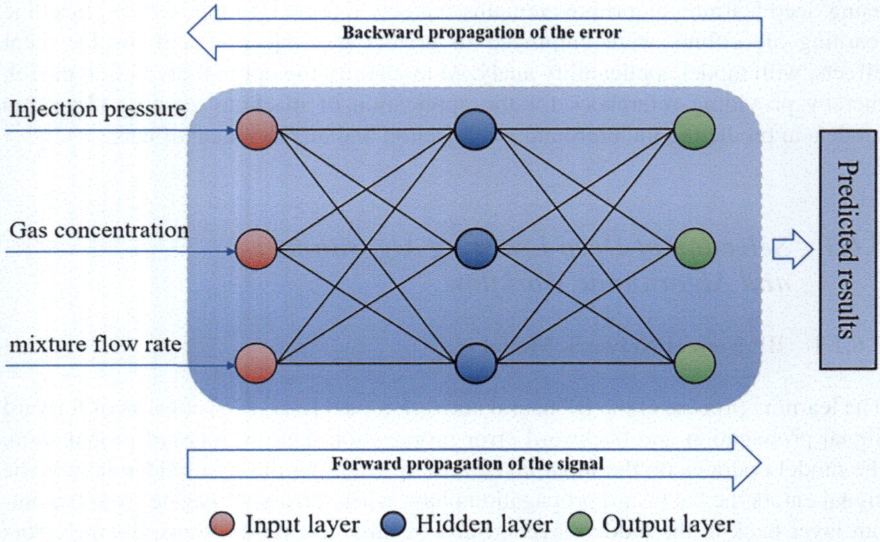

Fig. 5.15 Schematic diagram of BP neural model structure

$$H_j = f\left(\sum_{i=1}^{n} w_{ij}x_i - a_j\right) \quad j = 1,2,\ldots,m \tag{5.5}$$

$$f(x) = \frac{1}{1+e^{-x}} \tag{5.6}$$

(b) Output layer output calculation.

$$O_k = \sum_{i=1}^{m} w_{jk}H_j - b_k \quad k = 1,2,\ldots,l \tag{5.7}$$

(c) Error computation.

$$e_k = Y_k - O_k \quad k = 1,2,\ldots,l \tag{5.8}$$

(d) Weight updating.

$$w_{ij} = w_{ij} + \eta H_j\left(1+H_j\right)x(i)\sum_{k=1}^{l} w_{jk}e_{jk} \quad i = 1,2,\ldots,n; j = 1,2,\ldots,m \tag{5.9}$$

$$w_{jk} = w_{jk} + \eta H_j e_k \quad j = 1,2,\ldots,m; k = 1,2,\ldots,l \tag{5.10}$$

(e) Threshold updating for hidden and output layers.

$$a_j = a_j + \eta H_j \left(1 - H_j\right) \sum_{k=1}^{l} w_{jk} e_k \quad j = 1,2,\ldots,m \tag{5.11}$$

$$b_k = b_k + e_k \quad k = 1,2,\ldots,l \tag{5.12}$$

In the equation, x represents the input, H denotes the output of hidden layer computation, f stands for the selected activation function of the hidden layer, O indicates the input to output layer computation, e corresponds to the model prediction error, and η signifies the learning rate.

4. If the error e falls below the preset target value within the specified number of iterations, the predicted value is output; otherwise, step 3 is repeated.

Although the BP neural network is renowned for its simplicity, intuitiveness, and ease of implementation, it is prone to overfitting when the number of parameters increases [230]. Furthermore, its performance often significantly deteriorates when handling sequences with medium- to long-term dependencies, manifesting as reduced prediction accuracy and necessitating repeated tuning.

5.6.1.2 TCN Neural Network Model

The temporal convolutional network (TCN) is a neural network architecture commonly employed for sequence modeling tasks [231]. As a variant of the convolutional neural network (CNN) architecture, it is specifically designed to process variable-length sequences. The model was initially proposed by Lea et al. [232] based on CNNs and subsequently developed by Bai et al. [233]. The model structure is illustrated in Fig. 5.16. The TCN model incorporates three distinctive features: causal convolution, dilated convolution, and residual connection [234]. These features endow the model with a flexible receptive field, which can be adjusted by modifying the dilation factor for time series of arbitrary lengths, thereby mapping them to output sequences of equal length. This design effectively prevents leakage of future information in time-series data. Additionally, it mitigates the risk of gradient explosion or vanishing gradients that may arise as the network deepens.

The incorporation of causal convolution in the TCN model is pivotal for its capability to process time-series data, effectively segregate future information, and maintain the temporal causality of sequential data. For one-dimensional input data $[x_0, x_1, \ldots, x_t, x_{t+1}]$, the output y_t at time step t is exclusively influenced by preceding inputs after processing through a 1D convolutional kernel, as formalized in Eq. (5.13).

$$y_t = f\left(x_0, x_1, x_2, \ldots, x_t\right) \tag{5.13}$$

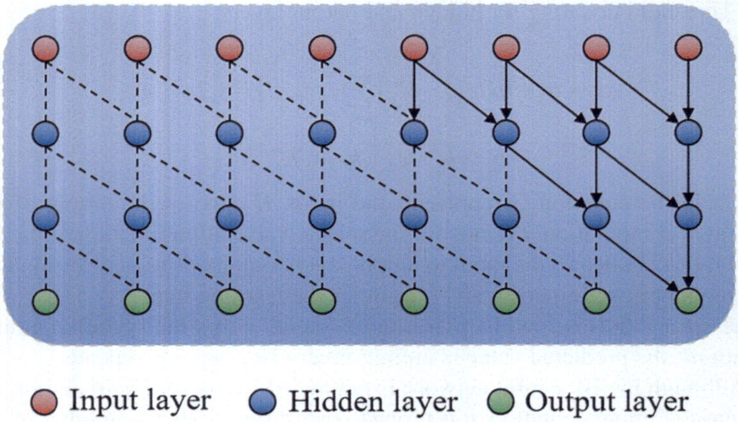

Fig. 5.16 Schematic diagram of TCN model structure

In the equation, y_t denotes the output of causal convolution at time step t, and f represents the one-dimensional convolutional kernel.

The introduction of dilated convolution in the temporal convolutional network (TCN) model aims to achieve exponential growth of the receptive field through dilation factors, thereby reducing the required convolution operations for long sequences. This design effectively simplifies the network architecture and decreases computational load. For one-dimensional input data $[x_0,x_1,\ldots,x_t,x_{t+1}]$, the output of the hidden layer at time step t after dilated convolution processing is given by Eq. (5.14).

$$h(t)=\sum_{i=0}^{kernel-1} f(\varepsilon)\cdot x_{t-\mu\cdot i} \tag{5.14}$$

In the equation, $h(t)$ represents the output of dilated convolution at time step t; $f(\varepsilon)$ denotes the convolutional kernel element at position ε; kernel indicates the size of the convolutional kernel, which is set to 3 in this study; $x_{t-m\cdot i}$ corresponds to the element of the sampled sequence after interval sampling; and μ is the dilation factor, specified as 1, 2, and 4 in this work.

During the information extraction process, the incorporation of residual connections serves to preserve critical initial features, thereby enhancing model stability and mitigating issues associated with vanishing or exploding gradients. The residual connection is mathematically expressed in Eq. (5.15).

$$y=F(x,W)+x \tag{5.15}$$

In the equation, y denotes the output of the residual connection; x represents the input data; and $F(x,W)$ corresponds to the extracted features after modeling.

Under conditions of limited sample data, the TCN model may exhibit overfitting [235], particularly when the network depth is substantial. Furthermore, hyperparameters such as convolutional kernel size and network depth significantly influence the efficacy of TCN. Consequently, meticulous parameter tuning is essential to achieve optimal results.

5.6.1.3 LSTM Neural Network Model

The long short-term memory (LSTM) network is a deep learning model designed for processing sequential data [236, 237]. As a variant of the recurrent neural network (RNN) architecture, it specifically addresses the prevalent gradient vanishing problem in conventional RNNs. The LSTM model comprises a series of memory cells capable of retaining information over extended periods, with each cell containing three gating mechanisms: an input gate, a forget gate, and an output gate. These gates collectively regulate the information flow into and out of the memory cell. The input gate determines which incoming information is relevant for storage; the forget gate decides which stored information should be discarded; and the output gate controls which information should be propagated from the memory cell. The model learns to predict subsequent elements in a sequence based on previously observed elements, while iteratively updating its parameters to minimize prediction errors. The architecture of the model is illustrated in Fig. 5.17.

The model information is transmitted according to the following sequence [238]: First, starting from the external state H_{t-1} at the previous time step and the input state x_t at the current time step, the three gate states are calculated based on Eqs. (5.16), (5.17), and (5.18), followed by obtaining the candidate state. After

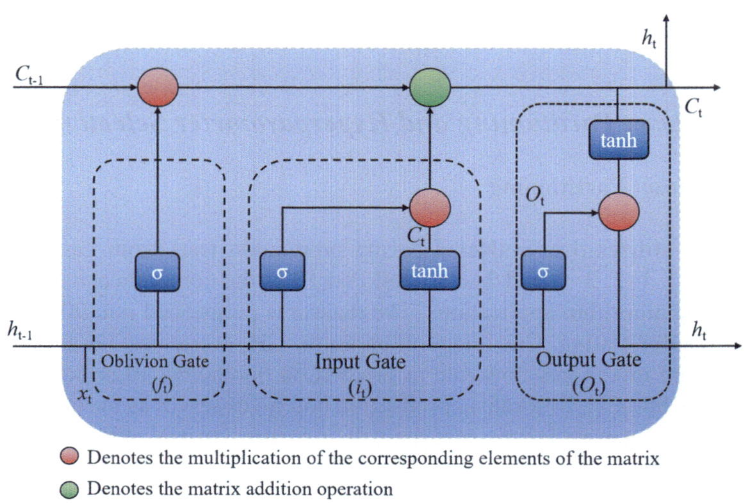

Fig. 5.17 Schematic diagram of LSTM model structure

completing the parameter regulation of the forget gate f_t and input gate i, the system executes the numerical update process for the memory cell C_t. Finally, the output gate o performs a nonlinear transformation on the updated memory cell and derives the output value of the external state H_t at the current time step according to the computational rule specified in Eq. (5.19).

$$i_t = \sigma\left(W_i \cdot \left[H_{t-1}, x_t\right] + b_i\right) \tag{5.16}$$

$$f_t = \sigma\left(W_f \cdot \left[H_{t-1}, x_t\right] + b_f\right) \tag{5.17}$$

$$o_t = \sigma\left(W_o \cdot \left[H_{t-1}, x_t\right] + b_o\right) \tag{5.18}$$

$$C_t = f_t \otimes C_{t-1} + i_t \otimes \tanh\left(W_c \cdot \left[H_{t-1}, x_t\right] + b_c\right) \tag{5.19}$$

$$H_t = o_t \otimes \tanh\left(C_t\right) \tag{5.20}$$

In the equation, $[W_i,b_i]$, $[W_f,b_f]$, $[W_o,b_o]$, and $[W_c,b_c]$ represent the weight matrices and biases from the hidden layer to the input gate, forget gate, output gate, and cell state, respectively. The symbols σ and tanh, as standard mathematical notations, correspond to the sigmoid activation function and hyperbolic tangent activation function, while the $\otimes$ operator strictly defines the element-wise multiplication operation between two tensors.

Although the LSTM model effectively addresses the issues of vanishing and exploding gradients, it tends to produce inaccurate predictions [239, 240]. This is primarily because achieving optimal prediction performance requires a substantial amount of data. When dealing with datasets exhibiting significant complexity, the LSTM model faces challenges in capturing subtle data variations.

5.6.2 Dataset Partitioning and Hyperparameter Selection

5.6.2.1 Dataset Partitioning

This section utilizes data collected during production tests from gas production boreholes 1–1, 1–2, 2–1, and 2–2 at Well Pad 10 as the comprehensive dataset for deep learning algorithm applications. The dataset is partitioned into 90% for training and 10% for testing. Gas injection pressure, CH_4 concentration at production boreholes, and mixed gas flow rate at production boreholes are selected as input parameters, while CH_4 flow rate at production boreholes serves as the output parameter. This parameter also represents the sole indicator for evaluating displacement effectiveness.

5.6.2.2 Model Hyperparameter Selection

The implementation was conducted using the Keras interface within the TensorFlow algorithm library, based on the Python integrated development environment PyCharm with Python 3.6. The process employed the Adam optimizer and mean squared error (MSE) loss function.

Model hyperparameters, which determine the training and performance of deep learning models, are predefined before training and typically remain constant throughout the training process. These parameters may encompass various settings, including learning rate, batch size, and activation functions, among others. Hyperparameters significantly influence model performance. Through careful hyperparameter tuning, model accuracy can be enhanced while addressing issues of overfitting or underfitting. Table 5.3 presents the selected hyperparameters for the models in this study. The prediction workflow diagrams for each model are illustrated in Fig. 5.18.

5.6.3 Selection of Evaluation Metrics

The mean absolute error (MAE) [241], root mean square error (RMSE) [242], and coefficient of determination (R^2) [243] were selected to evaluate the performance of deep learning models in predicting N_2 injection for enhanced coalbed methane drainage.

5.6.3.1 Mean Absolute Error

MAE represents the average absolute difference between predicted and actual values. It is calculated by taking the absolute value of the difference between each predicted value and its corresponding actual value, then averaging all these absolute differences. The expression is given by:

Table 5.3 Hyperparameter settings of each model

Model	Hyperparameter
TCN	Batch_size=20, Iteration=1500, Dense=1, Activation=linear, Dropout_rate=0.01, Optimizer name=adam, Loss function name=mae, Validation_split=0.1, Input(shape=(2,4)), windows=2, features=3
LSTM	Batch_size=20, Iteration=1500, Dense=1, Activation=linear, Dropout_rate=0.01, Optimizer name=adam, Loss function name=mae, Validation_split=0.1, Input(shape=(2,4)), windows=2, features=3
BP	Batch_size=len (train_data), Iteration=35,000, Learning_rate=0.2, gradient descent=Small batch gradient descent

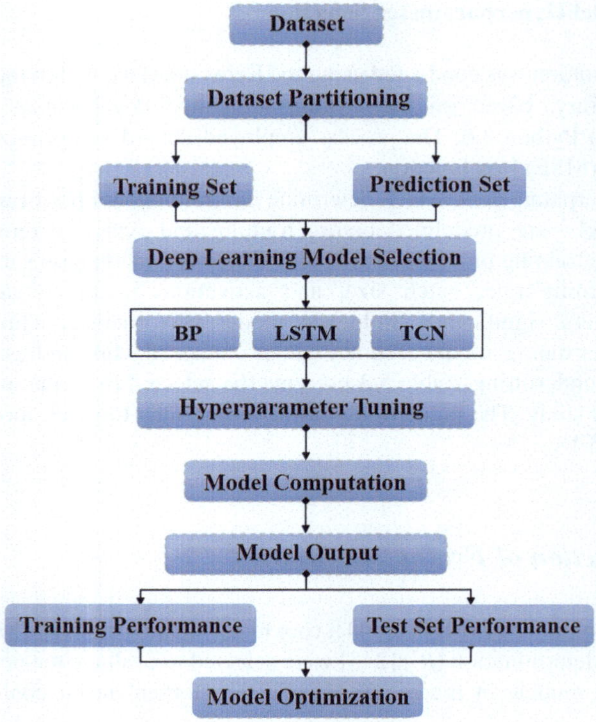

Fig. 5.18 Prediction flowchart

$$MAE = \frac{1}{n}\sum_{i=1}^{n}\left|predicted_i - actual_i\right| \tag{5.21}$$

5.6.3.2 Root Mean Square Error

The root mean square error (RMSE) serves as a prevalent metric for evaluating regression model performance, effectively characterizing the magnitude of average prediction errors. RMSE computation involves squaring the differences between predicted and actual values prior to extracting the square root of their average, thereby assigning greater weight to larger errors. The mathematical expression is formulated as:

$$\text{RMSE} = \sqrt{\frac{\sum_{i=1}^{n}\left(predicted_i - actual_i\right)^2}{n}} \tag{5.22}$$

5.6.3.3 Coefficient of Determination (R^2)

The coefficient of determination (R^2) is a statistical measure that quantifies the proportion of variance in the dependent variable explained by the independent variables in a regression model. An R^2 value of 1 indicates that all variability in the dependent variable is perfectly explained by the model's independent variables, whereas an R^2 value of 0 suggests no explanatory power. Higher R^2 values demonstrate better model fit to the observed data. The mathematical formulation is expressed as:

$$R^2 = 1 - \frac{\sum_{i=1}^{n} \left(y_i - \hat{y}_i\right)^2}{\sum_{i=1}^{n} \left(y_i - \hat{y}_i\right)^2}, \in [0,1] \tag{5.23}$$

5.6.4 *Training and Test Set Prediction Results*

5.6.4.1 Training Set Prediction Results

The prediction results of the training set represent the degree of fit of the predictive model to the training data, with the purpose of enabling the predictive model to learn the patterns and relationships within the dataset. Figure 5.19 shows the training set prediction results of each model. From Fig. 5.19a, it can be seen that the prediction results of all models are consistent with the overall trend of the actual values, and even the prediction results at oscillation points are satisfactory. From Fig. 5.19b–d, it is clearly visible that the actual values and predicted results of all models are well distributed on both sides of the $y = x$ accuracy line, indicating that each model can better capture the overall trend of the data and is unlikely to produce extreme prediction results.

Further quantitative evaluation of the model prediction results shows that for the TCN, LSTM, and BP neural network prediction models, the MAE values are: 0.01300, 0.01457, and 0.04596, respectively; the corresponding RMSE values are: 0.01713, 0.03753, and 0.08789; and the R^2 values are: 0.99818, 0.99125, and 0.95202. The training set prediction results preliminarily indicate that all models can effectively learn the basic patterns and relationships of the dataset, with a high degree of fit to the training set. In horizontal comparison, the TCN model provides the best prediction performance for the training set, followed by the LSTM model, while the BP model performs relatively poorly.

5.6.4.2 Test Set Prediction Results

Upon completion of model training, predictions conducted on an independent test set provide a robust evaluation of model performance on newly generated data. Figure 5.20 presents the predictive performance of all models on the test set.

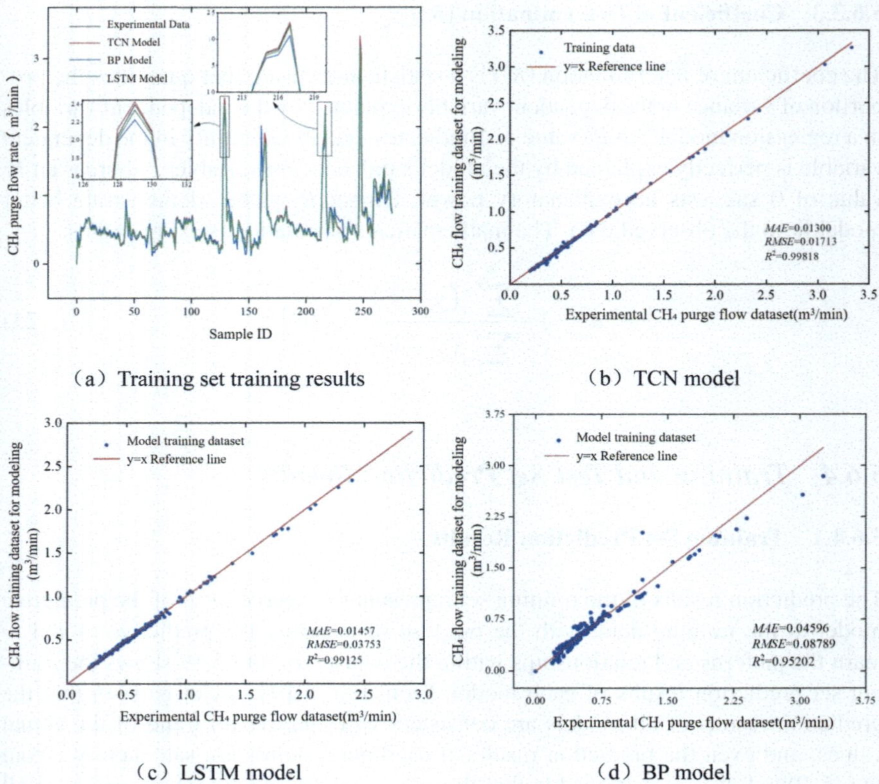

Fig. 5.19 Predictive performance of the model training sets. **(a)** Training set training results. **(b)** TCN model. **(c)** LSTM model. **(d)** BP model

Figure 5.20a demonstrates that the prediction results on the test set exhibit similar characteristics to those on the training set, with the predicted values generally following the overall trend of actual values. However, detailed examination of the magnified view reveals discrepancies between predicted and actual values across all models. Preliminary analysis indicates that the TCN model exhibits the smallest deviation from actual values.

Figure 5.20b–d presents the optimization results of three algorithms using the dataset of underground coal mine N_2 injection for CH_4 displacement. In each subfigure, the abscissa represents experimental data, while the ordinate denotes predicted data. A quantitative evaluation of the model predictions yields the following metrics: the TCN, LSTM, and BP neural network models demonstrate MAE values of 0.01109, 0.02236, and 0.02640, respectively; corresponding RMSE values of 0.02192, 0.03552, and 0.04044; and R^2 values of 0.99454, 0.98568, and 0.98099.

The results indicate that for the test set, the TCN model maintains superior performance, with predicted values clustering closely around the $y = x$ accuracy line. It

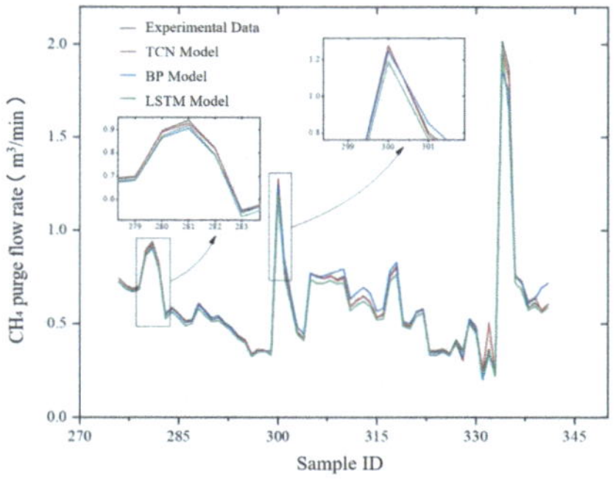

(a) Training set training results

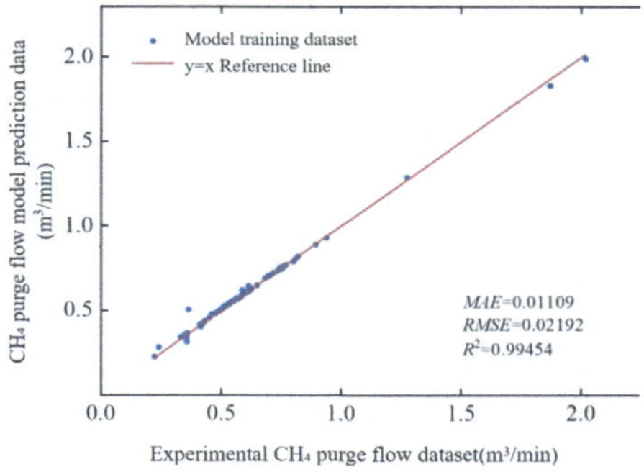

(b) TCN model

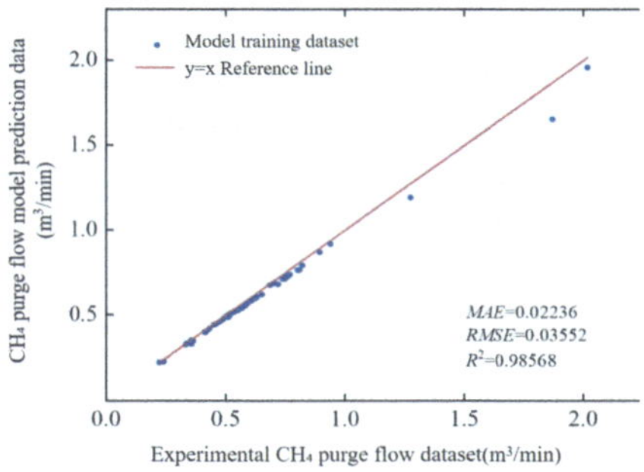

(c) LSTM model

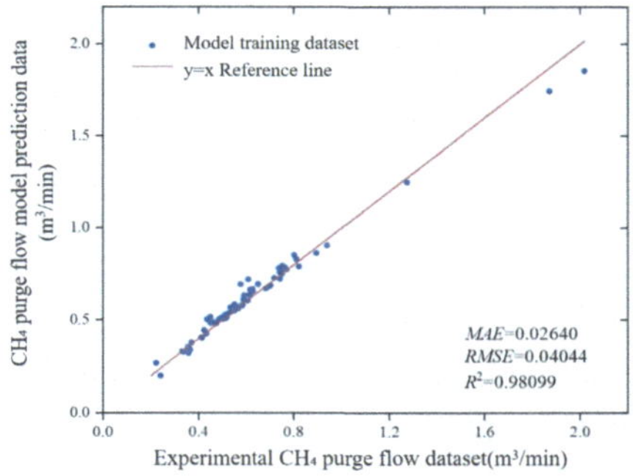

(d) BP model

Fig. 5.20 Predictive performance of the model test sets. (**a**) Training set training results. (**b**) TCN model. (**c**) LSTM model. (**d**) BP model

outperforms both LSTM and BP neural network algorithms across all evaluation metrics. The lower MAE and RMSE values, coupled with higher R^2, confirm that TCN predictions exhibit closer alignment with actual measurements. These findings demonstrate that the TCN model provides more accurate and reliable predictions for CH_4 pure flow rate during N_2 injection coalbed methane displacement operations.

5.6.5 Analysis of TCN Model Advantages and Applicability

5.6.5.1 Advantage Analysis of TCN Model

The prediction results in Sect. 5.6.4 demonstrate that all three models can effectively predict the variation in pure CH_4 flow rate during underground coal mine N_2 injection displacement operations, though with varying degrees of accuracy. Compared to LSTM and BP neural network models:

1. The TCN model exhibits relatively stable prediction performance on both training and test sets, delivering optimal results primarily due to its strong robustness against noisy data. The convolutional layers in TCN effectively filter out noise, whereas LSTM and BP neural networks may overfit to noise in input sequences. This is reflected in the TCN model's higher R^2 values.
2. The TCN model demonstrates superior parallel processing capability. While the BP neural network achieves satisfactory prediction performance ($R^2 > 0.95$ for both training and test sets on the N_2 injection displacement dataset), Table 5.2 shows that the TCN model requires only 1500 epochs to achieve highly accurate predictions, compared to 35,000 epochs for BP neural networks (23 times more). This efficiency stems from TCN's fully parallelizable operations across sequence steps, whereas LSTM and BP models are inherently sequential, limiting their scalability and computational efficiency.
3. For smaller datasets, the TCN model learns sequential features more effectively than LSTM and BP neural networks, thereby improving prediction accuracy and generalizability. Accurate predictions from limited sample data not only provide reliable estimates of pure CH_4 flow rates during N_2 injection displacement but also offer valuable theoretical guidance for predicting underground CH_4 extraction volumes, gas pressure, and concentration changes—all of significant practical importance.

In summary, test results confirm that TCN excels at handling nonlinear dynamic relationships and capturing long-term dependencies in irregular sequences. This is achieved through convolutional kernels that effectively identify patterns across different time scales. The parallel computing nature of TCN gives it superior performance over LSTM and BP neural networks when processing complex data.

5.6.5.2 Applicability Analysis of TCN Model

In practical coal mining engineering, significant variations exist in displacement outcomes due to differences in gas content, pressure, and reservoir structure within coal seams across mining areas with complex geological formations. A predictive model demonstrating consistently high accuracy when forecasting displacement results in diverse mining areas would possess greater applicability in real-world engineering scenarios.

As previously established, the TCN model exhibits exceptional accuracy in predicting CH_4 emissions during underground N_2 injection coalbed CH_4 displacement operations. To validate the model's applicability, we collected distinct datasets. The underground N_2 injection coalbed CH_4 experimental dataset comprises 151 samples sourced from Ref. [194]. Additionally, another dataset containing 152 samples was obtained from Ref. [27], featuring surface well N_2 injection coalbed CH_4 test results.

The model hyperparameters remain consistent with those detailed in Sect. 5.6.2, with the input layer retaining gas injection pressure, gas concentration, and mixed gas flow rate. Test results are presented in Figs. 5.21 and 5.22.

The TCN model consistently demonstrates accurate prediction results. For both surface and underground N_2 injection coalbed CH_4 tests, the deviations between predicted and actual values remain minimal across training and test sets. Predicted values closely cluster around the $y = x$ accuracy line, with prediction performance R^2 exceeding 0.97. This high precision confirms the TCN model's capability for accurate predictions in such datasets. The model effectively handles noise and various data variations, delivering optimal performance in predicting pure CH_4 flow rate changes during N_2 injection displacement processes. These characteristics establish the TCN model as an essential tool for accurately predicting pure CH_4 flow rate variations during underground gas injection displacement operations in coal mines.

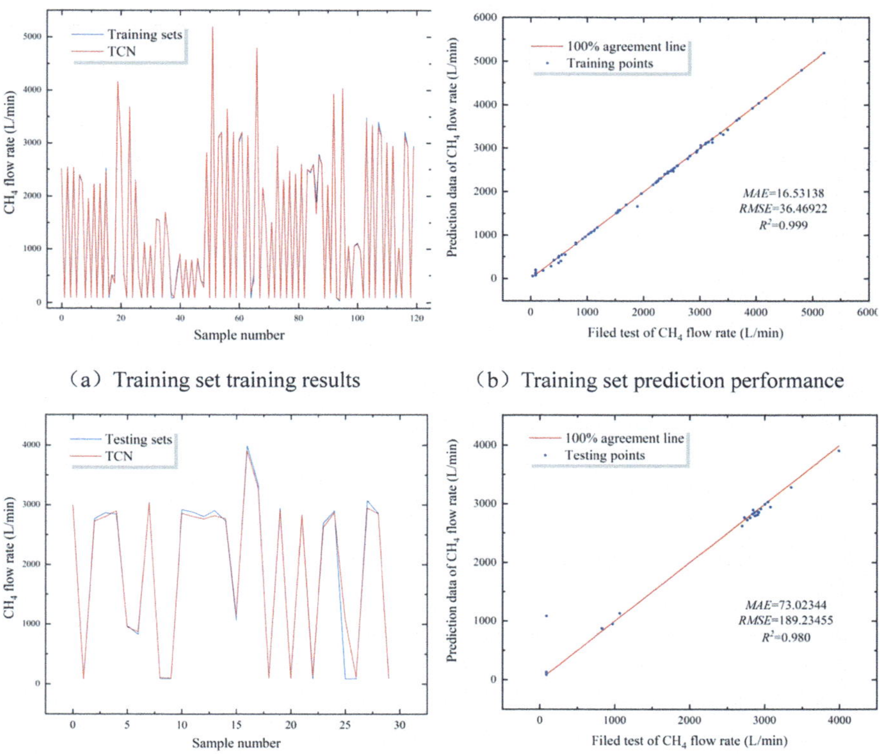

(a) Training set training results (b) Training set prediction performance

(c) Test set prediction results (d) Test set prediction performance

Fig. 5.21 N_2 injection in surface wells to replace CBM. (**a**) Training set training results. (**b**) Training set prediction performance. (**c**) Test set prediction results. (**d**) Test set prediction performance

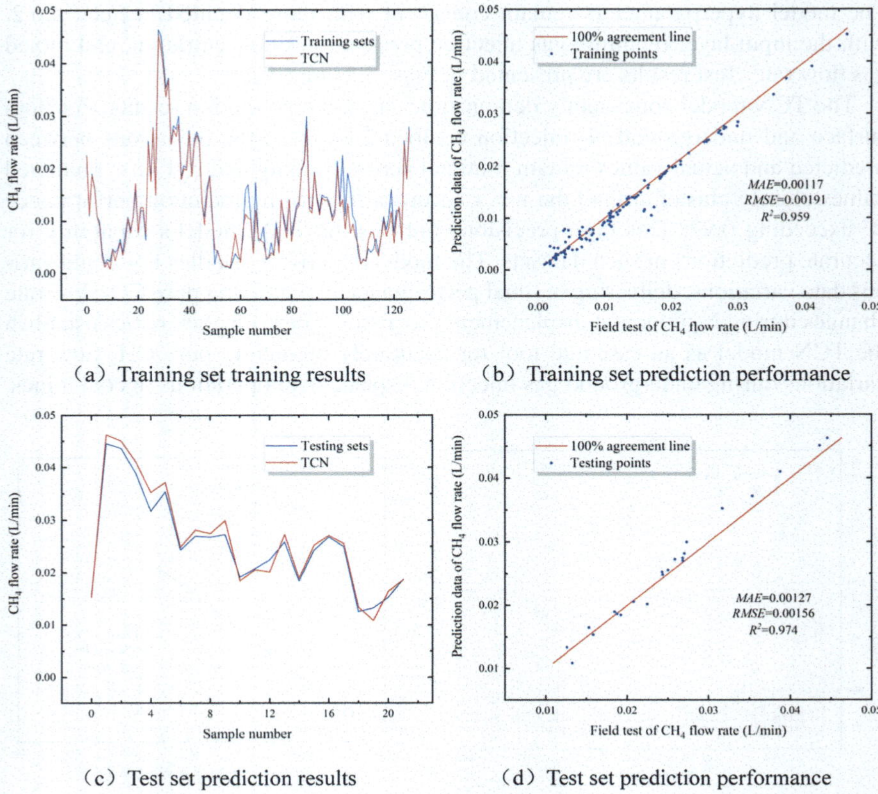

Fig. 5.22 Underground N_2 injection to replace CBM. (**a**) Training set training results. (**b**) Training set prediction performance. (**c**) Test set prediction results. (**d**) Test set prediction performance

5.6.6 Significance and Development Trends of Accurate Prediction for Nitrogen Injection Displacement of Coalbed Methane Pure Volume

Deep learning models demonstrate significant potential for development as they can process complex nonlinear relationships without requiring sophisticated mathematical derivations or idealized coal seam condition assumptions. These models adapt to diverse data types and applications, enabling optimization of N_2 injection displacement processes in ways unattainable by traditional mathematical models. The N_2 injection displacement technology effectively extracts CH_4 that would otherwise remain adsorbed under normal pressure conditions. Accurate prediction of pure CH_4 volume variations during N_2 injection displacement facilitates process optimization and proper timing of "gas injection-production" operations, ensuring maximum CH_4 recovery. This approach significantly reduces coalbed gas content, mitigates safety hazards such as coal and gas outbursts and gas explosions, and enhances coal mine safety production capacity. Moreover, optimized N_2 injection processes enable mines to minimize both the required N_2 volume for ideal production and the carbon footprint of coal mining operations, yielding substantial practical benefits for coalbed methane extraction optimization, mining safety improvement, and gas control cost reduction.

The primary objective of this study focuses on employing the TCN model to accurately predict pure CH_4 volume variations during N_2 injection displacement, thereby enhancing coalbed methane production while reducing gas content. However, several potential improvements merit consideration:

(a) Expanded Input Feature Dimensions: The current study only incorporates gas injection pressure, CH_4 concentration, and mixed gas flow rate as input layer parameters. Future research could improve prediction accuracy by incorporating additional dynamic parameters such as original coalbed gas pressure and content variations during N_2 injection displacement processes.
(b) Integration with Complementary Technologies: The TCN model's prediction accuracy and efficiency could be further enhanced through synergistic combinations with other techniques. For instance, machine learning-based data preprocessing methods could eliminate outliers and improve overall data quality.
(c) Hardware-Validated Implementation: Field testing through integration of the validated TCN model with sensor hardware would verify model reliability by identifying discrepancies between predictions and actual field data, thereby guiding further model optimization.

5.7 Chapter Summary

This chapter implemented underground N_2 injection for coalbed CH_4 displacement engineering applications at borehole sites $10^{\#}$, $13^{\#}$, and $16^{\#}$ along the belt entry of panel 151,105 in the $11–2^{\#}$ coal seam of China Coal Xinji Energy Group's Liuzhuang Mine, utilizing a self-developed "pressure boosting-gas storage-gas injection" experimental system. The study investigated the engineering characteristics of CH_4 dynamic transport under N_2 injection conditions, quantitatively analyzed the differential effects on CH_4 drainage between "gas injection displacement" versus "conventional drainage" methods and between "intermittent injection" versus "continuous injection" techniques and examined the influence of injection flow rate on CH_4 drainage efficiency. Based on the engineering application results, the intrinsic mechanisms of N_2-enhanced CH_4 drainage at engineering scale were elucidated. Additionally, preliminary exploration and model optimization were conducted for applying deep learning algorithms to predict gas injection displacement effects at engineering scale. The specific conclusions are as follows:

1. Injecting N_2 at pressures exceeding reservoir gas pressure effectively enhances the pressure gradient between fractures and production boreholes, increasing the

flow velocity of mixed gas within coal seam fractures. Following N_2 injection, the average mixed gas flow rates at production boreholes increased by factors of 2.56, 5.83, 2.97, and 2.20, respectively, compared to conventional drainage termination, while the average pure CH_4 flow rates increased by factors of 1.42, 2.66, 2.21, and 1.89 correspondingly.

2. Compared to conventional drainage, the underground N_2 injection displacement process significantly enhanced coalbed CH_4 emissions. Throughout the post-displacement phase, the measured pure CH_4 flow rates predominantly exceeded the conventional drainage prediction curve, with total CH_4 emissions increasing by 64.9%, 222.5%, 132.3%, and 149.1% relative to conventional drainage predictions. After N_2-enhanced drainage, the average coalbed gas content in the test area decreased notably by 1.151 m^3/t compared to pre-injection measurements. However, the substantially reduced spacing between injection and production boreholes (compared to surface well coalbed methane enhancement projects) made N_2 breakthrough time nearly negligible, consequently increasing CH_4 purification costs to some extent.
3. Compared to continuous injection, intermittent injection demonstrates more pronounced enhancement of CH_4 drainage from coal seams. Under continuous injection, the predicted cumulative CH_4 emissions from production boreholes during the period from injection cessation to test termination were 7.5, 2.19, 3.55, and 9.51 m^3, respectively. In contrast, intermittent injection yielded cumulative CH_4 emissions of 8.88, 17.46, 19.57, and 17.39 m^3 from corresponding boreholes. These results indicate that periodic N_2 injection not only enhances coalbed methane production but also improves gas hazard control by extracting greater volumes of CH_4. This approach achieves dual benefits: reducing injection volume for cost savings while simultaneously increasing CH_4 output for enhanced economic returns.
4. Analysis of injection flow rate effects on CH_4 drainage efficiency reveals that during initial injection phases, higher flow rates produce faster increases in pure CH_4 flow rates and higher peak values, though this effect's duration shortens with increasing flow rates. As displacement progresses, greater injection rates accelerate both the rate and magnitude of pure CH_4 flow decline. During later stages, differences in pure CH_4 flow rates across various injection rates gradually diminish. Final cumulative CH_4 emissions at manifold locations under 10, 15, and 20 m^3/h injection rates reached 123.3, 122.8, and 124.9 m^3, respectively, demonstrating minimal variation in total CH_4 recovery across different flow rates.
5. The underground coal mine N_2 injection-enhanced CH_4 drainage process can be divided into two distinct phases, each governed by different dominant mechanisms. During the initial phase, the pressure gradient drives the production of primarily original free-phase CH_4 from fractures, while the secondary phase is characterized by concentration gradient-driven production of CH_4 desorbed from originally adsorbed states.
6. In predicting CH_4 pure flow rates during gas injection and displacement processes, deep learning algorithm models (BP, LSTM, and TCN) demonstrate superior predictive performance. Across both training and testing datasets, the

approximation error rates of the three models follow the descending order BP > LSTM > TCN. Specifically, the TCN, LSTM, and BP neural network models exhibit MAEs of 0.01109, 0.02236, and 0.02640, respectively, on the training set, with corresponding RMSEs of 0.01713, 0.03753, and 0.08789 and R^2 values of 0.99818, 0.99125, and 0.95202. On the testing set, these models show MAEs of 0.02192, 0.03552, and 0.04044, RMSEs of 0.02192, 0.03552, and 0.04044, and R^2 values of 0.99454, 0.98568, and 0.98099. Consequently, the TCN model outperforms both LSTM and BP models across all quantitative evaluation metrics, demonstrating optimal performance in predicting CH_4 pure flow rates during N_2 injection displacement processes, while also exhibiting excellent applicability in other analogous field test results.

References

1. 李洪言，于淼，郑一鸣，等. 基于情景的2050年世界能源供需展望分析——基于《bp世界能源展望 (2023年版)》[J]. 天然气与石油， 2023， 41(5):131–137
2. Busch A, Gensterblum Y (2011) CBM and CO_2-ECBM related sorption processes in coal: a review [J]. Int J Coal Geol 87(2):49–71
3. Lin J, Ren T, Cheng Y et al (2019) Cyclic N_2 injection for enhanced coal seam gas recovery: a laboratory study [J]. Energy 188:116115
4. Du F, Wang K, Zhang X et al (2020) Experimental study of coal-gas outburst: insights from coal-rock structure, gas pressure and adsorptivity [J]. Nat Resour Res 29(4):1–13
5. 袁亮，薛生. 煤层瓦斯含量法确定保护层开采消突范围的技术及应用[J]. 煤炭学报，2014(9): 6
6. 翟成，李贤忠，李全贵. 煤层脉动水力压裂卸压增透技术研究与应用[J]. 煤炭学报，2011， 36(12): 6
7. 郝富昌，孙丽娟，左伟芹. 考虑流变特性的水力冲孔孔径变化规律及防堵孔技术[J]. 煤炭学报， 2016(6): 7
8. 卢义玉，丁红，葛兆龙，等. 空化水射流热效应影响煤体渗透率试验研究[J]. 岩土力学， 2014， 35(5):8
9. 林柏泉，吕有厂，李宝玉，等. 高压磨料射流割缝技术及其在防突工程中的应用[J]. 煤炭学报， 2007， 32(9): 5
10. 霍中刚. 二氧化碳致裂器深孔预裂爆破煤层增透新技术[J]. 煤炭科学技术， 2015，43(002): 80–83
11. Liu Z, Cheng Y, Wang Y et al (2019) Experimental investigation of CO_2 injection into coal seam reservoir at in-situ stress conditions for enhanced coalbed methane recovery [J]. Fuel 236:709–716
12. Silva DP, Ranjith P (2013) Advanced core flooding apparatus to estimate permeability and storage dynamics of CO 2 in large coal specimens [J]. Fuel 104:417–425
13. Lin J, Ren T, Wang G et al (2018) Experimental investigation of N_2 injection to enhance gas drainage in CO_2-rich low permeable seam [J]. Fuel 215:665–674
14. 宋鑫，舒龙勇，王斌，等. 低瓦斯赋存高强度开采煤层驱替促抽技术研究[J]. 采矿与安全工程学报， 2023， 40(4): 847–856
15. Mazzotti M, Pini R, Storti G (2009) Enhanced coalbed methane recovery [J]. J Supercrit Fluids 47(3):619–627
16. Pan Z, Connell LD (2009) Comparison of adsorption models in reservoir simulation of enhanced coalbed methane recovery and CO_2 sequestration in coal [J]. Int J Greenh Gas Con 3(1):77–89

G. Wang, H. Gong, *Dynamic Migration Characteristics of CH4 During Gas Injection Displacement in Coal Seams*,
https://doi.org/10.1007/978-981-95-8879-4

17. Kumar H, Elsworth D, Mathews J P et al (2014) Effect of CO_2 injection on heterogeneously permeable coalbed reservoirs[J]. Fuel 135:509–521
18. Shi JQ, Durucan S, Fujioka M (2008) A reservoir simulation study of CO_2 injection and N_2 flooding at the Ishikari coalfield CO_2 storage pilot project, Japan [J]. Int J Greenh Gas Con 2(1):47–57
19. Wong S, Law D, Deng X et al (2007) Enhanced coalbed methane and CO_2 storage in anthracitic coals—micro-pilot test at South Qinshui, Shanxi, China [J]. Int J Greenh Gas Con 1(2):215–222
20. Bergen FV, Krzystolik P, Wageningen NV et al (2009) Production of gas from coal seams in the Upper Silesian Coal Basin in Poland in the post-injection period of an ECBM pilot site [J]. Int J Coal Geol 77(1–2):175–187
21. Mazumder S, Wolf KH (2008) Differential swelling and permeability change of coal in response to CO_2 injection for ECBM [J]. Int J Coal Geol 74(2):123–138
22. Zhou F, Hussain F, Cinar Y (2013) Injecting pure N_2 and CO_2 to coal for enhanced coalbed methane: experimental observations and numerical simulation [J]. Int J Coal Geol 116(Complete):53–62
23. Gunter W (2005) CO_2 storage and enhanced methane production field testing at Fenn-Big Valley, Alberta, Canada, with application [C]. In: Greenhouse gas control technologies 7, vol I, pp 413–421
24. Durucan S, Shi JQ (2009) Improving the CO_2 well injectivity and enhanced coalbed methane production performance in coal seams [J]. Int J Coal Geol 77(1–2):214–221
25. Reeves S R. The Coal-Seq project: key results from field, laboratory, and modeling studies [C]. Greenhouse gas control technologies 7, 2005, II - Part 1: 1399–1403
26. Bustin AMM, Bustin RM, Chikatamarla L et al (2016) Learnings from a failed nitrogen enhanced coalbed methane pilot: Piceance Basin, Colorado [J]. Int J Coal Geol 165:64–75
27. Packham R, Connell L, Cinar Y et al (2012) Observations from an enhanced gas recovery field trial for coal mine gas management [J]. Int J Coal Geol 100:82–92
28. Wang H, Fu X, Jian K et al (2015) Changes in coal pore structure and permeability during N_2 injection [J]. J Nat Gas Sci Eng 27:1234–1241
29. Wang L, Wang Z, Li K et al (2015) Comparison of enhanced coalbed methane recovery by pure N_2 and CO_2 injection: experimental observations and numerical simulation [J]. J Nat Gas Sci Eng 23:363–372
30. 吴家全. 多组分气体混合物在多孔固体上吸附平衡研究[D]. 天津:天津大学， 2006
31. 杨其銮，王佑安. 煤屑瓦斯扩散理论及其应用 [J]. 煤炭学报， 1986(03): 89–96
32. 彭英明，邵先杰，李锋，等. 煤层气达西、非达西渗流理论和扩散理论的研究进展综述 [J]. 煤炭工程， 2019(6):6
33. Dubinin MM (1960) The potential theory of adsorption of gases and vapors for adsorbents with energetically nonuniform surfaces [J]. Chem Rev 60(2):235–241
34. Brunauner S, Emmett SPH, Teller EJ (1938) Adsorption of gases in multimolecular layers, J [J]. J Am Chem Soc 60(2):309–319
35. Langmuir I (1918) The adsorption of gases on plane surfaces of glass, mica and platinum [J]. J Am Chem Soc 40(9):1361–1403
36. 谢自立，郭坤敏. 微孔容积填充吸附理论的研究 [J]. 化工学报， 1995(4):452–457
37. 聂百胜，段三明. 煤吸附瓦斯的本质 [J]. 太原理工大学学报， 1998(4): 88–92
38. Qing-Ling Z (2008) Adsorption mechanism of different coal ranks under variable temperature and pressure conditions [J]. 中国矿业大学学报:英文版 18(3):6
39. Hao S, Chu W, Jiang Q et al (2014) Methane adsorption characteristics on coal surface above critical temperature through Dubinin–Astakhov model and Langmuir model [J]. Colloids Surf A Physicochem Eng Asp 444:104–113
40. 盛茂，李根生，陈立强，等. 页岩气超临界吸附机理分析及等温吸附模型的建立[J]. 煤炭学报， 2014， 39(S1): 179–183
41. APC, ASAM, ARLR et al (2015) Modeling the temperature dependence of supercritical gas adsorption on activated carbons, coals and shales [J]. Int J Coal Geol 138:113–126

42. 王飞. 煤的吸附解吸动力学特性及其在瓦斯参数快速测定中的应用[D]. 徐州:中国矿业大学， 2016
43. Jaroniec M, Deryło A, Marczewski A (1983) The Langmuir-Freundlich equation in adsorption from dilute solutions on solids [J]. ChemInform 14(4):393–397
44. Yelash LV, Kraska T (2001) A generic equation of state for the hard-sphere fluid incorporating the high density limit [J]. Phys Chem Chem Phys 3(15):3114–3118
45. Kuznetsov SV, Bobin VA (1980) Desorption kinetics during gas dynamical phenomena in collieries [J]. J Min Sci 16(1):49–55
46. Ettinger IL (1987) Solubility and diffusion of methane in coal strata [J]. Sov Min Sci 23(2):159–169
47. Levy JH, Day SJ, Killingley JS (1997) Methane capacities of Bowen Basin coals related to coal properties [J]. Fuel 76(9):813–819
48. Jian X, Guan P, Zhang W (2012) Carbon dioxide sorption and diffusion in coals: experimental investigation and modeling [J]. Sci China Earth Sci 55(4):633–643
49. Pone JDN, Halleck PM, Mathews JP (2009) Sorption capacity and sorption kinetic measurements of CO_2 and CH_4 in confined and unconfined bituminous coal [J]. Energy Fuel 23(5):4688–4695
50. Svabova M, Weishauptova Z, Pribyl O (2012) The effect of moisture on the sorption process of CO_2 on coal [J]. Fuel 92(1):187–196
51. Shi JQ, Durucan et al (2003) A bidisperse pore diffusion model for methane displacement desorption in coal by CO_2 injection [J]. Fuel 82:1219–1229
52. Cui X, Bustin RM, Dipple G (2004) Selective transport of CO_2, CH_4, and N2 in coals: insights from modeling of experimental gas adsorption data [J]. Fuel 83:293–303
53. Busch A, Gensterblum Y, Krooss BM et al (2004) Methane and carbon dioxide adsorption–diffusion experiments on coal: upscaling and modeling [J]. Int J Coal Geol 60(2/4):151–168
54. Siemons N, Wolf KHAA, Bruining J (2007) Interpretation of carbon dioxide diffusion behavior in coals [J]. Int J Coal Geol 72(3–4):315–324
55. Wang G, Ren T, Qi Q et al (2017) Determining the diffusion coefficient of gas diffusion in coal: development of numerical solution [J]. Fuel 196:47–58
56. 刘彦伟. 煤中瓦斯放散规律、机理与动力学模型研究[D]. 焦作:河南理工大学， 2011
57. Staib G, Sakurovs R, Gray EMA (2015) Dispersive diffusion of gases in coals. Part I: model development [J]. Fuel 143:612–619
58. Staib G, Sakurovs R, Gray EMA (2015) Dispersive diffusion of gases in coals. Part II: an assessment of previously proposed physical mechanisms of diffusion in coal [J]. Fuel 143:620–629
59. Wang G, Ren T, Qi Q et al (2017) Prediction of coalbed methane (CBM) production considering bidisperse diffusion: model development, experimental test, and numerical simulation [J]. Energy Fuel 31(6):5785–5797
60. 杨鑫. 低渗煤层高压注氮驱替强化抽采技术及应用研究[D]. 北京:中国矿业大学(北京)， 2019
61. 马丹，李檣，张吉雄等. 断层破碎带岩体孔隙结构表征与非线性渗流特性[J].煤炭学报， 2023， 48(2):666–677
62. 梁卫国，张倍宁，韩俊杰等. 超临界CO_2驱替煤层CH_4装置及试验研究[J]. 煤炭学报， 2014， 39(08):1511–1520
63. 梁卫国，吴迪，赵阳升. CO_2驱替煤层CH_4试验研究[J]. 岩石力学与工程学报， 2010， 29(04):665–673
64. 王立国. 注气驱替深部煤层CH_4实验及驱替后特征痕迹研究[D]. 徐州:中国矿业大学， 2013
65. 吴迪. 热力作用下CO_2驱替残留煤层CH_4渗流吸附解吸置换规律研究[D]. 葫芦岛:辽宁工程技术大学， 2013
66. 周军平，鲜学福，李晓红等. 吸附不同气体对煤岩渗透特性的影响[J]. 岩石力学与工程学报， 2010， 29(11): 2256–2262

67. 张美红，吴世跃，李川田. 煤系地层注入CO_2开采煤层气质交换的机理[J]. 煤炭学报，2013， 38(07): 1196–1200
68. Connell LD, Sander R, Pan Z et al (2011) History matching of enhanced coal bed methane laboratory core flood tests [J]. Int J Coal Geol 87(2):128–138
69. Kumar H, Elsworth D, Liu J et al (2015) Permeability evolution of propped artificial fractures in coal on injection of CO_2 [J]. J Pet Sci Eng 133:695–704
70. Ranathunga A, Perera M, Ranjith P et al (2017) Super-critical carbon dioxide flow behaviour in low rank coal: a meso-scale experimental study [J]. J CO_2 Util 20:1–13
71. 张子戌，刘高峰，张小东等. CH_4/CO_2不同浓度混合气体的吸附-解吸实验[J].煤炭学报， 2009， 34(04):551–555
72. 于洪观，范维唐，孙茂远等. 煤对CH_4/CO_2二元气体等温吸附特性及其预测[J].煤炭学报， 2005，(05):76–80
73. 周军平. CH_4、CO_2、N_2及其多元气体在煤层中的吸附-运移机理研究[D]. 重庆:重庆大学， 2010
74. 杨宏民，王兆丰，任子阳. 煤中二元气体竞争吸附与置换解吸的差异性及其置换规律[J]. 煤炭学报， 2015， 40(07):1550–1554
75. Clarkson C, Bustin R (2000) Binary gas adsorption/desorption isotherms: effect of moisture and coal composition upon carbon dioxide selectivity over methane [J]. Int J Coal Geol 42(4):241–271
76. Gruszkiewicz M, Naney M, Blencoe J et al (2008) Adsorption kinetics of CO_2, CH_4, and their equimolar mixture on coal from the Black Warrior Basin, West-Central Alabama [J]. Int J Coal Geol 77(1):23–33
77. 张登峰，崔永君，李松庚等. 甲烷及二氧化碳在不同煤阶煤内部的吸附扩散行为[J]. 煤炭学报， 2011， 36(10): 1693–1698
78. 易俊，姜永东，鲜学福. 煤层微孔中甲烷的简化双扩散数学模型[J]. 煤炭学报，2009， 34(03): 355–360
79. 聂百胜，郭勇义，吴世跃等. 煤中瓦斯扩散的理论模型及其解析解[J]. 中国矿业大学学报， 2001，(01): 21–24
80. 李志强，王司建，刘彦伟等. 基于动扩散系数新扩散模型的构造煤瓦斯扩散机理[J]. 中国矿业大学学报， 2015， 44(05): 836–842
81. 刘彦伟，刘明举. 粒度对软硬煤中瓦斯解吸扩散差异性的影响[J]. 煤炭学报， 2015，40(03): 579–587
82. 刘彦伟，魏建平，何志刚等. 温度对煤中瓦斯扩散动态过程的影响规律与机理[J]. 煤炭学报， 2013， 38(S1): 100–105
83. Charriere D, Pokryszka Z, Behra P (2010) Effect of pressure and temperature on diffusion of CO_2 and CH_4 into coal from the Lorraine basin (France) [J]. Int J Coal Geol 81(4):373–380
84. Pillalamarry M, Harpalani S, Liu S (2011) Gas diffusion behavior of coal and its impact on production from coalbed methane reservoirs [J]. Int J Coal Geol 86(4):342–348
85. 杨鑫，张俊英，王公达等. 瓦斯压力对瓦斯在煤中扩散影响的实验研究[J]. 中国矿业大学学报， 2019， 48(03): 503–510+519
86. Zhang H, Cao D (2016) Molecular simulation of displacement of shale gas by carbon dioxide at different geological depths [J]. Chem Eng Sci 156:121–127
87. Wu H, Chen A et al (2015) Molecular dynamics simulations about adsorption and displacement of methane in carbon nanochannels [J]. J Phys Chem C Nanomater Interfaces 119(24):13652–13657
88. 孙致学，闵成，张婉露等. CO_2/N_2二元气体对甲烷在煤中吸附影响的分子模拟研究[J]. 煤田地质与勘探， 2022， 50(03):127–136
89. Yang B, Hai-Fei L, Shu-Gang L et al (2021) Molecular simulation of N2 and CO_2 injection into a coal model containing adsorbed methane at different temperatures [J]. Energy 219:119686
90. 李树刚，白杨，林海飞等. N_2/CO_2注入压力对含瓦斯煤岩中甲烷解吸的影响[J]. 天然气工业， 2021， 41(03):80–89
91. 伦嘉云. 煤体纳米孔隙结构气体吸附特性研究[D]. 北京:中国矿业大学(北京)， 2020

92. 相建华. 不同煤级煤的结构特征及煤与CH_4、CO_2、H2O相互作用的分子模拟[D]. 太原:太原理工大学， 2013
93. 夏毅. 煤质结构及气化活性研究[D]. 贵阳:贵州大学， 2015
94. 王延斌，韩德馨，刘咸卫. 渤海湾盆地C-P煤有机组分^{13}CNMR研究[J]. 中国矿业大学学报， 1999，(01): 44–47
95. 王琦. 低阶褐煤热解过程的原位红外及拉曼光谱研究[D]. 大连:大连理工大学， 2016
96. 朱学栋，朱子彬. 红外光谱定量分析煤中脂肪碳和芳香碳[J]. 曲阜师范大学学报(自然科学版)， 2001，(04): 64–67
97. 韩雅婷. 不同煤级煤纳米力学性质及其对结构的响应[D]. 太原:太原理工大学， 2022
98. 程南南，石梦岩，侯泉林等. 拉曼光谱在煤大分子结构表征中的应用[J]. 煤炭学报，2023， 48(03): 1311–1324
99. 解迎芳. 煤结构及其在煤热解过程中的拉曼光谱研究[D]. 上海:上海大学， 2019
100. Wiser WH (1984) Conversion of bituminous coal to liquids and gases: chemistry and representative processes [M]. In: Magnetic Resonance. Springer, Dordrecht
101. 田力. CH_4、CO_2在烟煤结构模型中吸附的分子模拟研究[D]. 青岛:中国石油大学(华东)， 2016
102. Krevelen V, Chermin HA (1960) Chemical structure and properties of coal [J]. Fuel 33:79–87
103. Given PH (1960) The distribution of hydroxyl in coal and its relation to coal structure [J]. Fuel 39:147–158
104. 朱之培，高晋生. 煤化学[M]. 上海:上海科学技术出版社， 1984
105. Carlson AG (2002) Computer simulation of the molecular structure of bituminous coal [J]. Energy Fuel 6(6):771–778
106. 王文静. 煤中N_2/CO_2微孔填充特性实验研究[D]. 葫芦岛:辽宁工程技术大学， 2022
107. Liu F, Chu W, Sun W et al (2012) A DFT study of methane activation on graphite surfaces with vacancy defects [J]. J Nat Gas Chem 21(6):708–712
108. Kecheng Z, Peixue J, Zengmin L et al (2019) Molecular simulation of carbon dioxide and methane adsorption in shale organic nanopores [J]. Energy Fuel 33(3):1785–1796
109. Shen Z, Meng Z (2022) Enhancing the efficiency of coal bed methane recovery by injecting carbon dioxide based on an anthracite coal macromolecular model and simulation methods [J]. Energy Fuel 36(12):6329–6342
110. Chen C, Sun J, Zhang Y et al (2020) Adsorption characteristics of CH_4 and CO_2 in organic-inorganic slit pores [J]. Fuel 265:116969
111. Youzhi H, Lanfeng Y, Peichao L et al (2018) Molecular simulations of methane adsorption behavior in illite nanopores considering basal and edge surfaces [J]. Energy Fuel 32(4):4783–4796
112. 郝有志，卢德唐，贾晓天等. 页岩纳米孔隙内气水混合流动的分子动力学模拟 [J]. 中国科学:物理学， 2018， 48(09):2 67–276
113. 陈佳乐. 页岩气在粗糙纳米孔隙中的赋存特性和驱替机制研究 [D]. 无锡:江南大学， 2023
114. 王伟，方志明，李小春. 煤样尺度的二氧化碳驱替煤层气数值模拟[J]. 中国矿业，2019， 28(02):121–125
115. Lin J, Ren T, Wang G et al (2018) Simulation investigation of N_2 injection enhanced gas drainage: model development and identification of critical parameters [J]. J Nat Gas Sci Eng 55:30–41
116. 杨宏民，魏晨慧，王兆丰等. 基于多物理场耦合的井下注气驱替煤层甲烷的数值模拟[J]. 煤炭学报， 2010， 35(S1):109–114
117. 吴金涛，侯健，陆雪皎等. 注气驱替煤层气数值模拟[J]. 计算物理， 2014，31(06):681–689
118. 王兆丰，李炎涛，夏会辉等. 基于COMOSOL的顺层钻孔有效抽采半径的数值模拟[J]. 煤矿安全， 2012， 43(10):4–6
119. 丁智奔. 基于渗透率动态变化的注气驱替CH_4实验及数值模拟分析[D]. 焦作:河南理工大学， 2020

120. Reeves S, Oudinot A (2004) A technical and economic sensitivity study of enhanced coalbed methane recovery and carbon sequestration in coal [R]. DOE Topical Report March
121. Reeves S, Taillefert A, Pekot L, Clarkson C. The allison unit CO_2-ECBM Pilot: a reservoir modeling study [R]. DOE Topical Report March 2004
122. Reeves S. Coal-Seq Project Update: field studies of ECBM recovery/CO_2 Seq in Coalseams [C]. Presented at The 7th International Conference on Greenhouse Gas Control Technologies (GHGT-7) Kyoto
123. Reeves S, Taillefert AA. Reservoir modeling for the design of the RECOPOL CO_2 Sequestration Project, Poland [R]. DOE Topical Report June 2002
124. Mavor MJ, Gunter WD, Robinson JR. Alberta multiwell micro-pilot testing for CBM properties, enhanced methane recovery and CO_2 storage potential [C]. SPE Annual Technical Conference and Exhibition, 2004, Houston. SPE 90256
125. Van Bergen F, Pagnier H, Krzystolik P. Field experiment of CO_2-ECBM in the Upper Silesian Basin of Poland [C]. Proceedings of the 8th International Conference on Greenhouse Gas Control Technologies, Trondheim, Norway, June 19–22, 2006
126. Komaki H, KANSO (2004) CO_2-coal sequestration project in Japan [C]. In: Presentations from the 3th international forum on geologic sequestration of CO_2 in deep, unmineable coalbeds. Coal-Seq III, Baltimore, MD., March 25th-26th
127. Wong S, Law D, Deng X et al (2007) Enhanced coalbed methane and CO_2 storage in anthracitic coals—micro-pilot test at south Qinshui, Shanxi, China [J]. Greenh Gas Control 1(2):215–222
128. Robinson JR, Kadatz B, Wong S, Gunter WD, Feng SL, Fan ZQ. ECBM micro-pilot test in the anthracitic coals of the Qinshui basin, China: field results and preliminary analysis [C]. In: Proc., 3rd Intl. Workshop of Prospective Roles of CO_2 Sequestration in Coal Seams, Sapporo, Japan, Hokkaido U., October
129. 尹书郭. 页岩层中CO_2封存和CH_4开采的影响因素研究[D]. 武汉:武汉科技大学， 2023
130. Chaback JJ, Morgan WD, Yee D (1996) Sorption of nitrogen, methane, carbon dioxide and their mixtures on bituminous coals at in-situ conditions [J]. Fluid Phase Equilib 117(1–2):289–296
131. Vishal V et al (2015) Influence of sorption time in CO_2-ECBM process in Indian coals using coupled numerical simulation [J]. Fuel 139:51–58
132. 杨宏民. 井下注气驱替煤层甲烷机理及规律研究[D]. 焦作:河南理工大学， 2010
133. 方志明. 混合气体驱替煤层气技术的机理及试验研究[D]. 武汉:中国科学院研究生院(武汉岩土力学研究所)， 2009
134. 李元星. 连续与间歇注空气驱替煤层气机理及实验研究[D]. 太原:太原理工大学， 2017
135. Moore TA (2012) Coalbed methane: a review [J]. Int J Coal Geol 101:36–81
136. Hodot BB (1966) Coal and gas outburst [M]. Chemical Industry Press, Beijing
137. Walker PL (1975) Chemistry and physics of carbon [J]. J Mater Sci Technol 13(12):1066–1066
138. Gan H, Nandi S, Walker P (1972) Nature of the porosity in American coals [J]. Fuel 51(4):272–277
139. IUPAC (1994) Recommendations for the characterization of porous solids [J]. Pure Appl Chem 66:1739–1758
140. Chandra G, Lo PY, Hitchcock PB et al (1987) A convenient and novel route to bis (η-alkyne) platinum(0) and other platinum(0) complexes from Speier's hydrosilylation catalyst H2 [Pt C16] xH2O. X-ray structure of [Ptl(η-CH2-CHSiMe2) 2O (P-t-Su3)] [J]. Organometallics 6(1):191–192
141. 赵伟. 粉化煤体瓦斯快速扩散动力学机制及对突出煤岩的输运作用[D]. 徐州:中国矿业大学， 2018
142. 琚宜文. 构造煤结构演化与储层物性特征及其作用机理[D]. 徐州:中国矿业大学， 2003
143. 于兴河. 油气储层地质学基础[M]. 北京:石油工业出版社， 2009
144. 程远平. 煤矿瓦斯防治理论与工程应用[M]. 徐州:中国矿业大学出版社， 2010
145. 俞启香. 矿井瓦斯防治[M]. 徐州:中国矿业大学出版社， 1992

146. 秦勇，徐志伟，张井. 高煤级煤孔径结构的自然分类及其应用[J]. 煤炭学报，1995(3):266–271
147. Clarkson CR, Bustin RM (1999) The effect of pore structure and gas pressure upon the transport properties of coal: a laboratory and modeling study. 1. Isotherms and pore volume distributions [J]. Fuel 78(11):1333–1344
148. 韩青. 煤中纳米孔隙气体赋存与流动微观机理研究[D]. 太原:太原理工大学， 2023
149. 周娟. 页岩气吸附与CO_2驱替及封存机理的分子模拟研究[D]. 北京:清华大学， 2021
150. 李娜. 复合页岩中气体吸附扩散与驱替特性的分子模拟研究[D]. 青岛:中国石油大学(华东)， 2020
151. 刘艳侠，孙佳佳，高岩等. 关于Lennard-Jones型势函数参数的几种拟合方法[J]. 辽宁大学学报(自然科学版)， 2008，(03):206–209
152. 余昊. 页岩基质复杂孔网多尺度气体输运机理研究[D]. 合肥:中国科学技术大学， 2022
153. Verlet L (1967) Computer "experiments" on classical fluids. I. Thermodynamical properties of Lennard-Jones molecules [J]. Phys Rev 159(1):98
154. Van Gunsteren WF, Berendsen HJ (1988) A leap-frog algorithm for stochastic dynamics [J]. Mol Simul 1(3):173–185
155. 冯剑，刘洪来，胡英. 耗散粒子动力学的优化修正Velocity Verlet算法[J]. 化工学报，2006，(08):1841–1847
156. 辛鑫昕，撒占友，杨帅等. 基于分子模拟的微观结构特征对无烟煤吸附CH_4、CO_2影响规律对比研究[J]. 煤炭科学技术， 2024， 52(03):106–114
157. Metropolis N, Rosenbluth AW, Rosenbluth MN et al (1953) Equation of state calculations by fast computing machines [J]. J Chem Phys 21(6):1087–1092
158. 胡彪. 煤中多尺度孔隙结构的甲烷吸附行为特征及其微观影响机制[D]. 徐州:中国矿业大学， 2022
159. 赵豪田，宋党育，吉小峰. 煤中狭缝型纳米孔隙吸附甲烷分子模拟[C] 中国煤炭学会瓦斯地质专业委员会. 2017年全国瓦斯地质学术年会论文集. 河南理工大学资源环境学院；中原经济区煤层(页岩)气河南省协同创新中心； 2017:9
160. 郝琦. 煤的显微孔隙形态特征及其成因探讨[J]. 煤炭学报， 1987，(04):51–56+97-101
161. 唐书恒，张静平，吴敏杰. 腐泥煤孔隙结构特征研究[J]. 天然气地球科学， 2013，24(02):247–251
162. 胡羽. 煤氧化对CO_2/CH_4吸附影响机制的分子模拟研究[D]. 哈尔滨:哈尔滨工业大学， 2021
163. 孟召平，田永东，李国富. 煤层气开发地质学理论与方法[M]. 北京:科学出版社， 2010
164. Martin MG, Siepmann JI (1998) Transferable potentials for phase equilibria. 1. United-atom description of n-alkanes [J]. J Phys Chem B 102(14):2569–2577
165. 张敬茹. 页岩CO_2、CH_4动态吸附特性模拟研究[D]. 大连:大连理工大学， 2022
166. 朱苏阳. 煤层气的吸附—解吸机理及应用研究[D]. 成都:西南石油大学， 2020
167. 郭海军. 煤的双重孔隙结构等效特征及对其力学和渗透特性的影响机制[D]. 徐州:中国矿业大学， 2018
168. IUPAC (1994) Recommendations for the characterization of porous solids. Pure Appl Chem 66:1739–1758
169. Lin T, Junchen H, Davoud D et al (2021) New insight into absorption characteristics of CO_2 on the surface of calcite, dolomite, and magnesite [J]. Appl Surf Sci 540:148320
170. 于永江，张春会，王来贵. 超声波干扰提高煤层气抽放率的机理[J]. 辽宁工程技术大学学报(自然科学版)， 2008， 27(06):805–808
171. 李敏敏. 冲击荷载下结构异性煤体孔隙-裂隙响应特性研究[D]. 焦作:河南理工大学， 2024
172. Hardy J, Pomeau Y, De Pazzis O (1973) Time evolution of a two-dimensional model system. I. Invariant states and time correlation functions [J]. J Math Phys 14(12):1746–1759
173. 郭照立，郑楚光. 格子Boltzmann方法的原理及应用[M]. 北京:科学出版社， 2009
174. Qian YH, d'Humières D, Lallemand P (1992) Lattice BGK models for Navier-Stokes equation [J]. Europhys Lett 17(6):479

175. D'Humieres D (1994) Generalized lattice-Boltzmann equations [J]. Prog Astronaut Aeronaut 159:450
176. 刘洋. 基于格子Boltzmann方法的多孔介质微观渗吸机理研究[D]. 北京:中国地质大学，2022
177. 董波. 非混相驱替过程的格子Boltzmann模拟[D]. 大连:大连理工大学，2011
178. 贾思昊. 基于LBM的多孔介质中CO_2/咸水两相运移模拟研究[D]. 大庆:东北石油大学，2023
179. 赵建林. 页岩/致密油气藏微纳尺度流动模拟及渗流规律研究[D]. 青岛:中国石油大学(华东)，2018
180. 关钰. 注非烃气体开采煤层气解/吸附及扩散过程的LBM模拟[D]. 太原:太原理工大学，2017
181. Ladd AJ (1994) Numerical simulations of particulate suspensions via a discretized Boltzmann equation. Part 1. Theoretical foundation [J]. J Fluid Mech 271:285–309
182. Ladd AJ (1994) Numerical simulations of particulate suspensions via a discretized Boltzmann equation. Part 2. Numerical results [J]. J Fluid Mech 271:311–339
183. Zou QS, He XY (1997) On pressure and velocity boundary conditions for the lattice Boltzmann BGK model [J]. Phys Fluids 9(6):1591–1598
184. 郭照立，郑楚光，李青等. 流体动力学的格子Boltzmann方法[M]. 湖北科技出版社，2002
185. 曾彦. 基于格子Boltzmann方法的页岩气多尺度渗流模型研究[D]. 北京:中国石油大学(北京)，2018
186. Jian W, Yixiang G, Zhang S et al (2023) Pore-scale lattice Boltzmann simulation of CO_2-CH_4 displacement in shale matrix [J]. Energy 278:23–40
187. Chahbani M, Labidi J, Paris J (2002) Effect of mass transfer kinetics on the performance of adsorptive heat pump systems [J]. Appl Therm Eng 22(1):23–40
188. Takeshi S (2013) Implicit temperature-correction-based immersed-boundary thermal lattice Boltzmann method for the simulation of natural convection [J]. Phys Rev E. , Statistical, nonlinear, and soft matter physics 87(6):063304
189. 董武书，张跃. 单一裂隙渗流的格子Boltzmann数值模拟[J]. 煤炭技术，2021，40(01): 82–85
190. Zhou L, Qu ZG, Chen L, Tao WQ (2015) Lattice Boltzmann simulation of gas–solid adsorption processes at pore scale level [J]. J Comput Phys 300:800–813
191. 彭智高. 煤层气多尺度动态扩散-吸附及渗透率演化规律研究[D]. 北京:中国矿业大学(北京)，2018
192. 吴亚洲. 注气驱替煤层瓦斯扩散规律研究[D]. 北京:中国矿业大学(北京)，2023
193. Linstrom P, Mallard W (2022) NIST chemistry WebBook, NIST standard reference database number 69[R]. National Institute of Standards and Technology, Gaithersburg MD, p 20899
194. Xin Y, Gongda W, Feng D et al (2022) N_2 injection to enhance coal seam gas drainage (N_2-ECGD): insights from underground field trial investigation [J]. Energy 239:200–213
195. Yang B, Hai-Fei L, Shu-Gang L et al (2022) Experimental study on kinetic characteristics of gas diffusion in coal under nitrogen injection [J]. Energy 254:206–218
196. Gang B, Jun S, Shigen F et al (2024) Effect of CO_2 injection on the gas desorption and diffusion kinetics: an experimental study [J]. Energy 288:129921
197. Peng-fei J, Hai-fei L, Xiang-guo K et al (2023) Experimental study on enhanced coal seam gas extraction by uniform pressure/pulse pressure N2 injection [J]. Fuel 351:128988
198. Fang T, Wang M, Gao Y et al (2018) Enhanced oil recovery with CO_2/N_2 slug in low permeability reservoir: molecular dynamics simulation [J]. Chem Eng Sci 197:204–211
199. 白刚. 低渗煤层注CO_2置换驱替瓦斯机理及应用研究[D]. 葫芦岛:辽宁工程技术大学，2018
200. 冯朝阳. 煤层注氮气对甲烷的促排效应实验研究[D]. 焦作:河南理工大学，2016
201. 陈立伟. 煤层注气过程中置-驱转换机制及压力场效应研究[D]. 沈阳:东北大学，2019
202. 高子善. 高压驱替相与CH_4在煤体中的吸附解吸及扩散与渗流规律研究[D]. 徐州:中国矿业大学，2019

203. Khan IM, Yasmeen T, Khan IM et al (2016) Research progress in the development of natural gas as fuel for road vehicles: a bibliographic review (1991–2016) [J]. Renew Sust Energ Rev 66:702–741
204. Sharafian A, Talebian H, Blomerus P et al (2017) A review of liquefied natural gas refueling station designs [J]. Renew Sust Energ Rev 69:503–513
205. Yue XL, Bin BG, Jin Y et al (2021) Review on hydrate-based CH_4 separation from low-concentration coalbed methane in China [J]. Energy Fuel 35(10):8494–8509
206. Crank J (1975) The mathematics of diffusion [M]. Clarendon Press
207. Jian X, Guan P, Zhang W (2012) Carbon dioxide sorption and diffusion in coals: experimental investigation and modeling [J]. Sci China Earth Sci 556:33–43
208. švábová M, Weishauptová Z, Přibyl O (2012) The effect of moisture on the sorption process of CO_2 on coal [J]. Fuel 92(1):187–196
209. Pone JDN, Halleck PM, Mathews JP (2009) Sorption capacity and sorption kinetic measurements of CO_2 and CH_4 in confined and unconfined bituminous coal. Energy Fuel 23:4688–4695
210. Clarkson CR, Bustin RM (1999) The effect of pore structure and gas pressure upon the transport properties of coal: a laboratory and modeling study. 2. Adsorption rate modeling [J]. Fuel 78(11):1345–1362
211. Ruckenstein EVAS, Vaidyanathan AS, Youngquist GR (1971) Sorption by solids with bidisperse pore structures [J]. Chem Eng Sci 26(9):1305–1318
212. Wei XR, Wang GX, Massarotto P, Golding SD, Rudolph V (2007) A review on recent advances in the numerical simulation for coalbed-methane-recovery process [J]. SPE Reserv Eval Eng 10(06):657–666
213. Thomas NL, Windle AH (1982) A theory of case II diffusion [J]. Polymer 23(4):529–542
214. Berens AR, Hopfenberg HB (1978) Diffusion and relaxation in glassy polymer powders: 2. Separation of diffusion and relaxation parameters [J]. Polymer 19(5):489–496
215. Staib G, Sakurovs R, Gray AME (2013) A pressure and concentration dependence of CO_2 diffusion in two Australian bituminous coals [J]. Int J Coal Geol 116-117:106–116
216. Li Z, Liu D, Cai Y et al (2016) Investigation of methane diffusion in low-rank coals by a multiporous diffusion model [J]. J Nat Gas Sci Eng 33:97–107
217. Xu H, Tang D, Zhao J et al (2015) A new laboratory method for accurate measurement of the methane diffusion coefficient and its influencing factors in the coal matrix [J]. Fuel 158:239–247
218. Day S, Fry R, Sakurovs R (2012) Swelling of coal in carbon dioxide, methane and their mixtures [J]. Int J Coal Geol 93:40–48
219. Wei M, Liu J, Elsworth D et al (2019) Influence of gas adsorption induced non-uniform deformation on the evolution of coal permeability [J]. Int J Rock Mech Min Sci 114:71–78
220. Haoran G, Kai W, Gongda W et al (2021) Underground coal seam gas displacement by injecting nitrogen: field test and effect prediction [J]. Fuel 306:168–180
221. Satya Harpalani A, Schraufnagel RA (1990) Shrinkage of coal matrix with release of gas and its impact on permeability of coal [J]. Fuel 69(5):551–556
222. Wang K, Gong H, Wang G et al (2024) N_2 injection to enhance gas drainage in low-permeability coal seam: a field test and the application of deep learning algorithms [J]. Energy 290:86–99
223. Vishal V, Singh T, Ranjith P (2015) Influence of sorption time in CO_2-ECBM process in Indian coals using coupled numerical simulation [J]. Fuel 139:51–58
224. Xiaoliang Z, Wenhao L, Lei Z, Sheng X (2023) Quantitative evaluation of the indexes contribution to coal and gas outburst prediction based on machine learning [J]. Fuel 338:127389
225. Ibrahim AF (2022) Application of various machine learning techniques in predicting coal wettability for CO_2 sequestration purpose [J]. Int J Coal Geol 252:103951
226. Deng J, Lei C, Xiao Y, Cao K, Ma L, Wang W et al (2018) Determination and prediction on "three zones" of coal spontaneous combustion in a gob of fully mechanized caving face [J]. Fuel 211:158–170

227. Liang W, Wang G, Ning X, Zhang J, Li Y, Jiang C et al (2020) Application of bp neural network to the prediction of coal ash melting characteristic temperature [J]. Fuel 260:116324
228. Karacan CÖ (2009) Forecasting gob gas venthole production performances using intelligent computing methods for optimum methane control in longwall coal mines [J]. Int J Coal Geol 79(4):131–144
229. Chen H, Wang Y, Zuo M, Zhang C, Jia N, Liu X et al (2022) A new prediction model of CO_2 diffusion coefficient in crude oil under reservoir conditions based on BP neural network [J]. Energy 239:122286
230. Liu Q, Shang J, Wang J, Niu W, Qiao W (2023) Evaluation and prediction of the safety management efficiency of coal enterprises based on a DEA-BP neural network [J]. Resour Policy 83:103611
231. Liao J, Hu J, Yan F, Chen P, Zhu L, Zhou Q et al (2023) A comparative investigation of advanced machine learning methods for predicting transient emission characteristic of diesel engine [J]. Fuel 350:128767
232. Lea C, Flynn MD, Vidal R et al (2017) Temporal convolutional networks for action segmentation and detection [C]. In: Proceedings of the IEEE conference on computer vision and pattern recognition. IEEE, Honolulu, Hawaii, USA, pp 156–165
233. Chen Z, Zhu Z, Jiang H, Sun S (2020) Estimating daily reference evapotranspiration based on limited meteorological data using deep learning and classical machine learning methods [J]. J Hydrol 591:125286
234. Cheng W, Wang Y, Peng Z, Ren X, Shuai Y, Zang S et al (2021) High-efficiency chaotic time series prediction based on time convolution neural network [J]. Chaos, Solitons Fractals 152:111304
235. Zhang K, Cao H, Thé J, Yu H (2022) A hybrid model for multi-step coal price forecasting using decomposition technique and deep learning algorithms [J]. Appl Energy 306:118011
236. Mahmoodzadeh A, Mohammadi M, Ghafoor Salim S, Farid Hama Ali H, Hashim Ibrahim H, Nariman Abdulhamid S et al (2022) Machine learning techniques to predict rock strength parameters [J]. Rock Mech Rock Eng 55(3):1721–1741
237. Mahdaviara M, Sharifi M, Bakhshian S, Shokri N (2022) Prediction of spontaneous imbibition in porous media using deep and ensemble learning techniques [J]. Fuel 329:125349
238. Mirsalari SA, Nazari N, Sinaei S, Salehi ME, Daneshtalab M (2022) FaCT-LSTM: fast and compact ternary architecture for LSTM recurrent neural networks [C]. IEEE Des Test 39(3):45–53
239. Sen N, Pengfei J, Yang X et al (2023) Prediction of CO concentration in different conditions based on Gaussian-TCN [J]. Sensors Actuators B Chem 376:133010
240. Xue G, Song J, Kong X, Pan Y, Qi C, Li H (2019) Prediction of natural gas consumption for City-level DHS based on attention GRU: a case study for a northern Chinese city [C]. IEEE Access 7:85–99
241. Frías-Paredes L, Mallor F, Gastón-Romeo M, León T (2018) Dynamic mean absolute error as new measure for assessing forecasting errors [J]. Energy Convers Manag 162:176–188
242. Karunasingha DSK (2022) Root mean square error or mean absolute error? Use their ratio as well [J]. Inf Sci 585:609–629
243. Li B, Qiu J (2023) Displacement prediction of open-pit mine slope based on SSA-ELM [J]. Front Earth Sci 11:166–188

The manufacturer's authorised representative in the EU is Springer Nature Customer Service Centre GmbH, Europaplatz 3, 69115 Heidelberg, Germany. If you have any concerns regarding our products, please contact ProductSafety@springernature.com

Printed and bound by CPI Group (UK) Ltd, Croydon, CR0 4YY
21/09/2026
02220012-0002